T0143174

The Gaia Hypothesis

science.culture   *A series edited by Adrian Johns*

Other science.culture series titles available:

The Scientific Revolution, by Steven Shapin (1996)
Putting Science in Its Place, by David N. Livingstone (2003)
Human-Built World, by Thomas P. Hughes (2004)
The Intelligibility of Nature, by Peter Dear (2006)
Everyday Technology, by David Arnold (2013)

MICHAEL RUSE

the

GAIA

hypothesis

science on a
pagan planet

The University of Chicago Press
Chicago and London

Michael Ruse is the Lucyle T. Werkmeister Professor of Philosophy and director of the Program in the History and Philosophy of Science at Florida State University.

The University of Chicago Press, Chicago 60637
The University of Chicago Press, Ltd., London

© 2013 by The University of Chicago
All rights reserved. Published 2013.

Printed in the United States of America

22 21 20 19 18 17 16 15 14 13 1 2 3 4 5

ISBN-13: 978-0-226-73170-4 (cloth)
ISBN-13: 978-0-226-06039-2 (e-book)

Library of Congress Cataloging-in-Publication Data

Ruse, Michael.
    The Gaia hypothesis : science on a pagan planet / Michael Ruse
        pages. cm. — (Science.culture)
    Includes bibliographical references and index.
    ISBN 978-0-226-73170-4 (cloth : alk. paper) — ISBN 978-0-226-06039-2
(e-book) 1. Gaia hypothesis. 2. Geobiology—Philosophy. I. Title. II. Series:
Science.culture.
    QH331.R8785 2013
    571—dc23
                                                            2013009939

♾ This paper meets the requirements of ANSI/NISO Z39.48–1992
(Permanence of Paper).

For Lizzie, with love

Contents

Preface

Much has been written on the Gaia controversy, both in print and increasingly on the Internet. James Lovelock's autobiography, *Homage to Gaia*, will surely take its place among the classics. Why, then, should we bother with yet another account? Primarily because, although this is a book on Gaia, it is not really a book about Gaia. It is rather a philosophical and historical meditation on the nature of science itself, one that uses Gaia as its focus and as a tool to explore broadly important questions. The aim is to show how today's thinking about empirical questions is deeply influenced by the past. Beliefs and, more important, differences, do not just spring up from nowhere. The way we think now is a heritage from the ways that people thought years ago. Moreover, it results not only from the ways that scientists thought, but from the ways in which culture molded and informed the thinking of us all. This is not to say that there are no deep and rapid changes in science specifically and culture generally. Of course there are, and part of the discussion is intended to show some of the most important changes and how and why they occurred. But it is also intended to show that, although working scientists often pride themselves on not reading anything older than a decade or so, those of us who are interested in how and why science functions would be foolish to follow suit. Nor do I mean to say that only culture counts and the facts go for naught. That is just silly. We shall see again and again that the empirical evidence drives claims and counterclaims. Science is a complex mixture of the empirical and the theoretical, the discovered and the created, and to think that science stands apart from the rest of life, pure and disinterested, is just as silly. Readers must judge for themselves whether I have been successful in my aims; however, I can say that in writing this book I have frequently come to see in a new light some old fact or belief that I thought was given and in no need of further interpretation. I hope that this will be your experience also. Most important, I have great admiration for my two principals. If I cannot convince you that the ideas of the people about whom I write deserve attention and understanding, then indeed I have failed you and them.

I speak in my subtitle of Earth as a "pagan planet." In one sense, such usage hardly demands comment. The great Greek philosophers, where we start our history of the idea of a living planet, obviously had little interest in the religious beliefs of illiterate tribesmen to the south of them, and did not know of Jesus Christ and His redeeming mission here on Earth, or of Muhammad and his life and meaning. Understood in this context, as something that stands outside the Abrahamic religions, by definition the birth of Gaia was the birth of a pagan idea. Today, as we shall learn, among the most enthusiastic of Gaia supporters are those who call themselves Pagans or neo-Pagans—I capitalize to distinguish them from the past—and often they look back to the Greeks as their inspiration. But I intend a little more by the term, namely, that we are talking of something—our home, the planet Earth—that has life, that has value, in its own right. It is significant that although, as we shall see, there have been and still are Christians who accept the Gaia hypothesis, there has often been tension (especially for Protestants) between Christian commitment and acceptance of Gaia. For Christians, most notably for those who take the sacred scriptures as the only basis for the true religion, only God has value, and all else derives from Him. It is very much the opposing idea, the extended sense of Earth as something with intrinsic value, that interests me. One might easily say that atomism is as much a pagan idea as Gaia, and yet because in itself atomism does not contain the same value commitments as Gaia, the belief does not raise quite the same issues and passions. Understand, therefore, that I speak of Earth as a "pagan planet" precisely to highlight its vibrancy, its life, and its value that stems from this.

From the beginning, I have been and remain in the debt of many people. I start with the staff of the *Chronicle of Higher Education*, who asked me to review some books on the Gaia idea, and I continue with my editor and friend at the University of Chicago Press, Karen Merikangas Darling, who saw that there was the kernel of a book in my review, and who has encouraged me, read earlier versions of this book, and generally been all one could want in a press editor. Adrian Johns, the general academic editor of the series within which this book appears, is an old friend, and he has been supportive and critical in just the measures that an author needs. The same goes for the referees used by the press. They were understanding and penetrating in the needed proportions and, thanks to them, the book is much better than it might have been. The same thanks go to Nicholas Murray, who did the copyediting, and Martin Young, who was my illustrator.

As always, my academic friends have been there to read, to argue, and (thank goodness) to laugh both with me and, at times, at me. I am particularly grateful to my colleague Fritz Davis; to David Sepkoski and Mark Borrello, who came to a mini-conference at Florida State University and roughed me up over an earlier version; to Joe Cain, who found my style a bit too "chatty" (who doesn't?); and above all to my friend and former student John Beatty, who put his fingers on at least three serious faults, and who, a mere week before I submitted the final manuscript, had me slashing away at the text as I dropped many bits that I found interesting but were irrelevant to my main story.

I am very grateful to the people whom I interviewed. I am a tad disappointed that Oberon Zell-Ravenheart did not insist on my speaking to him "sky clad." Perhaps the thought of me without any clothes on was too much for even the most dedicated of Pagans. I must also thank George Handley for filling me in on the Mormons. As always, I am in debt to my family, especially to my wife Lizzie. (I still think it was a little mean of her to refuse to allow me to display my figurine of pregnant Gaia that I bore back proudly from my trip to California.) I am in debt also to my families back in England. First, the family in which I grew when my mother was living and we were all very committed Quakers. Much that Lovelock says and thinks resonates with me from long past. Second, the family I had after my mother's death, my father and my stepmother, with whom I spent my teenage years. I never shared their dedication to the ideas of Rudolf Steiner, but I hope the reader will see that my skepticism about his metaphysics is infused with sympathetic understanding and appreciation of his good influence on this world. To my surprise, I feel somewhat the same way about the Pagans. Finally, I am grateful to my home institution, Florida State University, not just for the time for research but also for the funds provided by my Lucyle T. Werkmeister Professorship.

A Note on
Interviews and Other Sources

The parenthetical abbreviations in the following source notes are used to cite these sources in the text. I conducted the following oral interviews:

Lynn Margulis (LM): December 8, 2010
Oberon Zell-Ravenheart (OZ): January 12, 2011
James Lovelock (JL): January 18, 2011
Andrew Watson (AW): January 23, 2011
Timothy Lenton (TL): January 24, 2011

In addition I have used the British Library, *An Oral History of British Science*, interviews (2010) of James Lovelock (BL); the David Suzuki (filmed) interview of James Lovelock, copyright 2002 (DS); an "online chat" with Lovelock conducted by the English newspaper the *Guardian* on September 29, 2000 (OC); "Gaia at 20: A Conversation with Lynn Margulis and James Lovelock" for a course given in the 1990s by Lynn Margulis entitled "Environmental Evolution" (EE); the Lynn Margulis Papers in the Library of Congress (LC); Joseph Cain's (University College London, Department of Science and Technology Studies) interviews (2001) with Robert E. Sloan (JC); Special Collections Research Center, University of Chicago (UC); the Rachel Carson Papers at Yale University (YU); and the Rachel Carson Papers at Connecticut College (CC). The letter from William Hamilton to James Lovelock is dated January 29, 1997, and was kindly copied for me by Tim Lenton. I am indebted to Christine A. Hamilton for permission to quote from it. Finally, the opinion of Herbert Spencer expressed by Edward O. Wilson dates back to 1982, when I first met Wilson and expressed shock and horror that he not only had a picture of Spencer on his wall but displayed it more prominently than that of Charles Darwin.

INTRODUCTION

"That's one small step for a man, one giant leap for mankind." Those are the famous words spoken by astronaut Neil Armstrong on Sunday, July 20, 1969, when he became the first man to step on the moon. It was the climax of the project begun by President John F. Kennedy when he declared on May 25, 1961, that the United States would beat Russia at its own game: "I believe that this nation should commit itself to achieving the goal, before this decade is out, of landing a man on the moon and returning him safely to the Earth." It was and still is by any measure a phenomenal technological achievement. To blast off from Earth, to travel to and circle the moon, to drop down a little craft to the surface, to walk about in an environment with no air or anything else that we humans need to survive, and then to return safely to Earth was not cheap, and it was not easy. It was almost certainly not the wisest way to explore the universe beyond our home. But it was done and done entirely successfully.

In the decade of the 1960s, the United States showed the world that it was the mightiest nation the globe has ever known—rich, powerful, knowledgeable, able to perform miracles. It was a time when social engineering in the form of the Civil Rights Act of 1964 was thought to be as powerful and long-lasting as physical engineering. But it was also a troubled decade with clear signs of difficulties to come. It was the decade of three, dreadful, hope-destroying assassinations: of President John F. Kennedy on November 22, 1963; of Martin Luther King, the leader of the march

1

to racial equality, on April 4, 1968; and then of the president's brother, Attorney General Robert F. Kennedy, on July 6, 1968. It was the decade when America increasingly involved itself in the civil war in Vietnam, committing huge numbers of troops and massive amounts of armory to the cause of the South as it battled the communist North. And it was also the decade when college campuses erupted, as students defied their elders and teachers and argued for more autonomy and a say, not just in the running of the universities but in the running of the country. It was a time when old notions of authority, status, and dignity stood for nothing. "Up against the wall, motherfucker," said one student rebel to the president of Columbia University, in the name of "liberation." But as with the trip to the moon, which was both part of the Cold War and yet a marvel of human ingenuity and understanding, higher education in the sixties was not all of one kind. Universities exploded in number and size during this time, making education more readily available to many. Many new positions and new departments were opened, and new kinds of people (including women) could engage in and benefit from what they offered. New ideas were, if not always welcome, at least allowed their chance to survive and flourish, particularly because of many new outlets for publishing. Thomas Kuhn, with his powerful notion of a "paradigm," showed us that science is a complex human activity, and that one should be wary of naive acceptance of the reductionistic, mechanistic philosophy that was perceived to lie behind so much of the thinking of the bureaucratic forces running the country.

To fill out the picture, the 1960s was the decade when this uneasy face-off between the established power of the older generation, backed by and enthusiastic about science and technology, and the rebellious doubt of the younger generation regarding the course of the nation and its authorities' enthusiasms led more and more people to explore new ways of making sense of existence, new dimensions of thought and action. Matters are rarely as simple and straightforward as the surface suggests. Overnight, the advent of birth-control pills changed sexual attitudes and behaviors as women were suddenly freed from the fear of unwanted pregnancy. Yet obviously, in its way, "the pill" was a triumph of the very technology that was being berated. One work that became standard reading for every teenager, William Golding's *Lord of the Flies* (fifteen thousand copies were sold in the United States in 1960 and more than half a million in 1962) is deeply rooted in the venerable doctrine of original sin. There was continuity and there was change. We see this very clearly in questions to do with ultimate meaning and practice. In the West, America has always been distinctive in its deeply religious foundation and nature. But the tensions of the times, whether they were rooted in the Cold War between the United States and Russia or in the rejection of the status quo and the search for a new order of things, led to explorations, developments, and innovations in unanticipated directions. On the right, reflecting the move of many Americans (particularly in the South) from traditional political bases to those offering comfort and protection against radical social changes, there was the rise of so-called Young Earth Creationism, which argued for a literal interpretation of the Bible—six thousand years since the beginning of the universe, six literal days of creation, a universal deluge shortly thereafter. Published in 1961, *Genesis Flood*, by biblical scholar John C. Whitcomb and hydraulic engineer Henry M. Morris, was the defining text. Its dispensational framework screamed the tensions of the times. The Flood was the end of the first period of Earth history, and Armageddon (with its images of nuclear warfare) will be the last. Are you ready? The Lord will come like a "thief in the night." Forget attempts to create paradise here on Earth and prepare for end times. On the left, also thinking in segments of time and history, many proclaimed our entry into the astrologically determined "Age of Aquarius." There was the obsession with Eastern religions, perhaps best reflected in popular culture by the friendship of the Beatles with the Maharishi Mahesh Yogi, deviser of Transcendental Meditation. But just as some went geographically outward to find their new metaphysics, some went historically backward to find their new metaphysics. There was a fascination with ancient mysteries and movements, with more basic, more Earth-centered creeds, often (fitting in with the spirit of the times) less patriarchal and more female-sensitive and also less technological and more organic or ecologically friendly. Completing the circle, the bible of all on this side of things was *Silent Spring*, published in 1962 by the powerful science writer Rachel Carson. She showed how a frenzied reliance on technology and science had led to the destruction of the environment—that our home was tainted and spoiled, unfit for us and our children, and crying for healing, for new, warmer ways of thinking and acting.

The Gaia Hypothesis: Science on a Pagan Planet tells a story that comes out of the 1960s, a story that reflects all of the beliefs and enthusiasms and tensions of that decade. It is a story that carries the themes through to the present, showing how the various ideas developed, changed, and matured, and sometimes withered. There are different lines, but they are not isolated, because they twist back and forth and entwine in some ways before diverging again. It is a story primarily but not exclusively about America. Britain in particular has a major contribution to make. That is no surprise. For all of the jokes about two countries separated by a common language, there is much cultural overlap, and that was true back then. The British adored Kennedy and the group around him, who represented such a break from the staid 1950s—the old war hero Dwight Eisenhower in the United States and the equally old Harold Macmillan in the United Kingdom. Similar social changes were happening. The number of university places doubled, thanks to the founding of new institutions in places like Sussex and Warwick. The Beatles, of course, were British, and for all that the old country is less intoxicated by religion than the new, some of the most influential movements had strong British links.

Although this is a story that comes out of the 1960s, it is not a story that began in the 1960s. Any evolutionist will tell you that the secret to the present is to be found in the past, and this holds as much in the realm of ideas as in the realm of organisms. In succeeding chapters, we dig back into the distant past. The exploration is fascinating in its own right, but always it is a story with an eye to future events and developments. The aim is not at all to show that we are wiser than those who went before, but to show that it is only in context that full understanding can emerge. The final chapters of analysis, when we return to the present era, will furnish the proof.

1

THE GAIA HYPOTHESIS

The moment of inspiration—the epiphany, one might say—came to English scientist James Lovelock one afternoon in September 1965. He was in California, working for NASA (the space agency), worrying about the composition of the atmosphere on Mars as opposed to that on Earth. The former is very different from the latter, the rich mixture within which we all live and that is so vital to our well-being. What could be the reason for the difference, or, more precisely, what could be the causes here that make our atmosphere a medium so far from the sterile equilibrium that we find on the Red Planet? "As Pasteur and others have said, 'Chance favours the prepared mind.' My mind was well prepared emotionally and scientifically and it dawned on me that somehow life was regulating climate as well as chemistry. Suddenly the image of the Earth as a living organism able to regulate its temperature and chemistry at a comfortable steady state emerged in my mind. At such moments, there is no time or place for such niceties as the qualification 'of course it is not alive—it merely behaves as if it were'" (Lovelock 2000, 253–54).

Lovelock tells a good and polished story. Was it actually this road-to-Damascus experience? There was an insight, although whether he had the full conception all at once is a little hazy. Perhaps it had to develop and mature. What is clear is that when he was back home in England, he was ready to start sharing his convictions—"I was already beginning to look on the Earth as an organism, or if not an organism, as a self-regulating system" (JL).

A crucial influence was none other than William Golding, author of *Lord of the Flies* and, in 1983, winner of the Nobel Prize for literature. He and Lovelock were neighbors in a small village and good friends. "When I first discussed it with Bill Golding, we went into it in considerable depth" (JL). The novelist was entranced by the idea; in fact, it was Golding who came up with the name Gaia, the Greek goddess of Earth. Yet, things did not really start to catch fire until Lovelock met and began collaboration with the American scientist Lynn Margulis. Apparently they first met at a meeting in 1968, but it was not until 1970 that they struck up a serious correspondence on the subject. They got together and started collaborating sometime toward the end of 1971. But me no buts. Earth is alive. It is an organism, it really is!

DRAMATIS PERSONAE

Who is Jim Lovelock, and who was Lynn Margulis (1938–2011)? Above all, in the circles of serious science, they are highly respected for their positive achievements (Turney 2003). He is a Fellow of the Royal Society of London, and she was a member of the (American) National Academy of Sciences, and no one begrudges them these honors. What I have to say in this book becomes a lot less interesting if one does not keep this fact firmly in mind. Lovelock (2000) tells us that he was born to a lower-middle-class family in England, just after the First World War, in 1919. (He claims to be the result of incautious celebration on Armistice Night, November 11, 1918!) He went to grammar school (the stream of publicly financed English secondary education reserved for bright pupils), and then, after a year or two of working for an industrial chemist, he got his undergraduate degree in chemistry. During the Second World War, he went to work for the government on practical issues such as the spread of the germs for the common cold—no small matter for bomber crews flying at high altitudes and wearing oxygen masks.

This was the beginning of twenty years of work on and around the boundaries of medicine and related areas of interest and importance. In retrospect, some of Lovelock's work seems to verge on the bizarre. For instance, he developed the technique for freezing and then resuscitating small mammals (a major concern for those wanting to preserve blood and other body parts). It became apparent early on that Lovelock had a real genius—and I use this term literally and deliberately—for instrument making: he was often able to make incredibly sensitive machines from war surplus and similar collections of junk. This did not come out of nowhere. From his earliest childhood, Lovelock was obsessed with science—he read and reread the science-fiction stories of the great writers, such as Jules Verne and H. G. Wells—but it was always science of a practical turn, the science of machinery. He would haunt the Science Museum (in South Kensington), awed and fascinated by the wonderful contraptions—steam engines, pumps, and the like, the life blood of the Industrial Revolution. The possibility of making and playing with machines drove him forward. His tendency to be somewhat of a loner—"I'm a little bit of an individualistic person" (BL)—led to his hobby becoming an obsession. By his own admission, referring to a sensible, all-weather coat that has become for the British a symbol of the socially inept, interest-absorbed outsider (e.g., train spotters), Lovelock became an "anorak of the first order" (BL).

What rescued him from obscurity was a warm and embracing personality, along with increasing recognition by others that his skills were leading to highly desirable products. The self-described "nerd" became a swan. Lovelock's most brilliant invention was a mechanism for detecting chemicals at infinitesimally small levels. The electron capture detector (ECD) is so precise that, to use Lovelock's example, it can record in Britain within a week or two the effects of emptying a bottle of solvent on a cloth in Japan. A man with such talents naturally attracted attention. He and his family spent several years in the United States while he worked at universities, and he found willing sponsors in both government agencies and private industry. So successful was Lovelock that he was able to quit his formal job and do freelance work, depending on his ability to produce things for organizations that needed his products and could pay well. Lovelock prides himself on this independence and frequently speaks scathingly of university hierarchies and (even more so) of granting agencies. Just as one suspects that there are many atheistic scientists who thank God for the Galileo affair, something they cite as proof of the awful nature of organized religion, so one suspects that Lovelock likewise thanks God for the foolish referee who derided one of his grant applications on the grounds that what he proposed was impossible. Like all sensible grant applicants, Lovelock had done enough of the work that he was already able to do the supposedly impossible—a fact that he still, some forty years later, reiterates with glee in almost every conversation. Something he is a little more reticent about—given that he was raised a Quaker and declared himself as a conscientious objector at the beginning of the Second World War—is the fact that defense establishments have gladly provided significant and regular funding for his production of sensitive instruments of detection.

Lovelock is not just a very clever scientist, he is an interesting man, with numerous fertile ideas and (as we shall see) a real talent for communicating with both general readers and specialists. He prides himself on his ability to move across boundaries: "I'm somewhat of a polymath. I feel at home in all branches of science" (BL). He was not fazed by those who thought it was daring, perhaps presumptuous, of an industrial chemist to propose a massive hypothesis about the nature of the whole planet on which we live. Although we shall learn later about the fundamental differences between the two Gaia enthusiasts, much that is true of Lovelock—especially his determination to push ideas because he thought them right rather than fashionable—applies as well to Lynn Margulis (Brockman 1995; Margulis and Sagan 1997). Twenty years younger than Lovelock, she was born and raised in Chicago, attending the University of Chicago at a ridiculously young age (fourteen), where she enrolled in the Great Books program. The heart of this educational program is working through uncut versions of the great classics of the West, such as Plato's *Republic* and Dante's *Inferno*, and her experience was surely a major factor in her refusal to be cowed by any authority or naysayer. She had learned to tackle head on the greatest minds of our civilization, and lesser mortals held no terrors. At nineteen, she married the man who was to become the best-known scientist in America, Carl Sagan, then an up-and-coming astronomer. She followed him first to Wisconsin (Madison) and then to the Berkeley campus of the University of California. Although she had two children and raised them pretty much unaided (the union with Sagan soon unraveled and then broke), Margulis enrolled in biology programs at both Wisconsin and Berkeley. Her first job after completing her PhD was at Boston University, where she stayed for many years before moving to the University of Massachusetts at Amherst to work in the Department of Geosciences.

In 1967, Lynn Sagan (as she was then) published in the *Journal of Theoretical Biology* a paper that had been rejected fifteen times. In "On the Origin of Mitosing Eukaryotic Cells" she argued that eukaryotic cells, that is, the complex cells with nuclei enclosing the chromosomes that carry (most of) an organism's genes (today understood to be lengthy molecules of ribonucleic acid), did not form de novo but are the results of symbiosis between more primitive cells, the prokaryotes (which have no nuclei and hence have the genes riding free). In particular, Margulis argued that some of the cell parts (organelles) of the eukaryotes, specifically including the mitochondria (the power plants that supply energy) and the plastids (particularly the chloroplasts that perform photosynthesis in plants), started life as free-existing, independent prokaryotes that were engulfed by other prokaryotes and (rather than dissolving) kept their own integrity and from then on contributed to the whole, that is, to the prokaryotes (now on their way to becoming eukaryotes) that incorporated them. Margulis (to use the name of her second husband, by which she was later known, even though that union also came to an end) was not the first to endorse "endosymbiotic theory," but at the time she published, it was ridiculed as unnecessary and improbable. Nothing if not persistent, Margulis followed her paper in 1970 with a detailed, book-length treatment of the topic (*The Origin of Eukaryotic Cells*), and slowly but surely the tide of opinion started to swing her way. The definitive evidence came in the 1980s, when gene sequencing had reached the level of sophistication that allowed comparisons between the nucleic acids found in organelles and those of various, promising, free-standing prokaryotes. The pertinent molecules were found to be virtually identical. Margulis was vindicated.

This was not to be the last time that neither praise nor condemnation could sway Lynn Margulis regarding a topic about which she had made up her mind. Her career was marked by controversies and the taking of unpopular positions. Like Jim Lovelock, she did not hesitate to take her case (or cases) to the public, and she wrote a number of books (several coauthored with her son, Dorion Sagan) that were specifically directed to the nonspecialist. With regard to the Gaia hypothesis, this determination and the ability to switch levels of discourse were important for both Lovelock and Margulis. Although it is true that back in 1970 Margulis had not yet gained the full respect of the scientific community, Lovelock was well known, and hence the collaborators expected that their ideas would be received at least respectfully, if critically. However, from the first they had trouble even publishing in professional journals. They were invited to a distinguished scientific conference to talk on the topic, but found, to their chagrin, that they "were not there as serious scientists but more as entertainment" (Lovelock 2000, 262). Fellow scientists did not want to discuss the ideas. They rejected Gaia "with that same certainty that the religious have when they reject the views of a rational atheist. They could not prove us wrong but they were sure in their hearts that we were" (263).

This did lead to what one might describe as a "teaser," for on the basis of his after-dinner talk, Lovelock published a short letter to the editor on Gaia in the journal *Atmospheric Environment*, but it was more a staking of claim than a detailed exposition and defense of the idea (Lovelock 1972). Fortunately, Lovelock and Margulis were not without resources. By the 1970s, old marital wounds had healed somewhat, and the encouragement and support of Carl Sagan was invaluable. He was the editor of the journal

Icarus, and it was here that—after a rejection by *Science* (Clarke 2012)—the Gaia hypothesis got its first prominent outing, and it is to this hypothesis that we now turn. Bear in mind that here and throughout the book I use the term *Gaia* to mean Lovelock's hypothesis; I use different words for the ideas of others, however close they are to Lovelock's. In this chapter my focus is on the basic early claims and reactions. There have been changes that are important for our overall discussion, but addressing them must be deferred.

WHAT IS GAIA, AND WHY IS IT NECESSARY?

Start with an indubitable but truly amazing fact. Since the formation of our solar system more than four billion years ago, because of the way in which the sun burns itself up, the energy it emits has been increasing over time. And not by some trivial amount. There could well have been a threefold increase over the years since the system began. Yet the surface temperature on Earth has remained almost constant, varying at most within a 10° Celsius band around today's mean (fig. 1). That this is just chance, given that the existent temperature is just about perfect for terrestrial surface life, is simply "unbelievable." Natural theologians of course would invoke the deity, but for scientists this is not an option. A naturalistic clue surely lies in the fact that, compared to that of other planets, Earth's atmosphere is different—very different. It stands out in all sorts of ways—with respect to its acidity, its composition, its temperature (after discounting differences stemming from distances from the sun), and much more. Moreover, there is solid evidence that this anomalous atmosphere is not something new. It has persisted over vast geological time periods. But how and why? Lovelock and Margulis opened their discussion by going straight to the heart of the matter. Using the term *homeostasis,* meaning "balance" or "equilibrium," they wrote, "We believe that these properties of the terrestrial atmosphere are evidence for homeostasis on a planetary scale." And just as the stability is essential for the well-being of organisms, so they postulated that these organisms

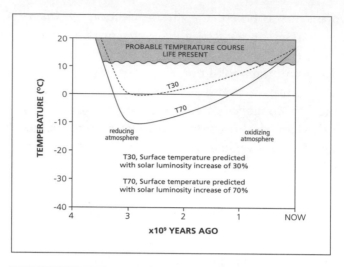

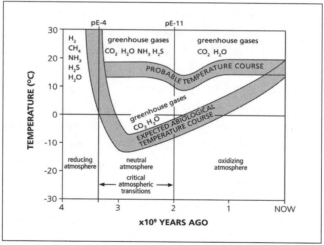

Figure 1. A. Temperature history of the earth: abiological prediction from solar luminosity. B. Temperature history of the earth: probable history derived from the fossil record. In a reducing atmosphere, oxygen is taken out; conversely, in an oxidizing atmosphere, oxygen is put in. The pE number measures reducing/oxidizing ability; a lower figure indicates reduction, and a higher figure indicates oxidation. (Reprinted, with slight modifications for clarity, from *Icarus* 21 [1974]: 471–89, L. Margulis and J. E. Lovelock, "Biological Modulation of the Earth's Atmosphere," figs. 2a and 2b, 475.)

themselves play a positive role in maintaining the stability. Bluntly, they stated that the "purpose of this paper is to develop the concept that the earth's atmosphere is actively maintained and regulated by life on the surface, that is, by the biosphere" (Margulis and Lovelock 1974, 471).

There is nothing like a good name to get a good idea off to a good start, so, having noted that the ancient Greeks used the term *Gaia* to refer to the "great communal being" made up of "all creatures on the earth, animals and plants, including man," Lovelock and Margulis clothed themselves in the veneration of antiquity by saying that "in deference to the ancient Greek tradition," they would "refer to the controlled atmosphere-biosphere as 'Gaia'" (Margulis and Lovelock 1974, 471). And so one asks: What work is Gaia (or, in the authors' words, the "Gaia hypothesis") going to do? We start (as did Lovelock when he started thinking on the subject) with the atmosphere. Venus has massive amounts of carbon dioxide, comparatively little nitrogen, and no oxygen at all. Mars has little carbon dioxide and virtually no nitrogen or oxygen. Earth, as is well known, has about 20% oxygen, 80% nitrogen, and traces of carbon dioxide. What's going on here? If you considered only the inorganic world, you would make no progress. However, things start to change when you factor in life on Earth, the biota. Using the term *cybernetics* for the study of systems where feedback mechanisms—meaning by *feedback* cases where the end product (the effect) swings around and affects the initial input (the cause) and thus controls or regulates the working of the whole—Earth is just such a system. We find a number of feedback mechanisms, such as we find in the control of a room's temperature by a thermostat and, even more pertinently, in the human body's regulation of its temperature through sweating and shivering. Indeed, the authors wrote, "We suspect that the earth's control systems follow a similar complex pattern more comparable to the temperature control in individual organisms than to man-made models" (474).

This leads to their statement of the Gaia hypothesis: "We conclude from the fact that the temperature and certain other environmental conditions on the earth have not altered very much from what is an optimum for life on the surface, that life must actively maintain these conditions" (Margulis and Lovelock 1974, 475). Life needs the conditions, so life sets about making the conditions. This is no mere a priori deduction, for we know for a fact that life has been involved, especially in altering the atmosphere and bringing us to the present state of affairs. The prokaryotes especially should not be ignored. They were working flat out before the eukaryotes (found in multicellular organisms) appeared, and, in any case, "nearly all the chemical transformations that large organisms perform microbes can do as well" (476). Certainly, we can put specific phenomena down to specific organisms; for instance, it is surely the case that oxygen was produced (about two billion years ago) by the photosynthetic activity of blue-green algae. We find similar explanations for other things, such as the alkalinity of the earth's surface. In all, "we believe that it is fruitful to assume that temperature, gas composition, and alkalinity have been actively modulated by organisms, especially microorganisms" (479).

How does the control take place? There are several causal systems affecting different parts of the final, balanced product. The surface temperature of Earth is in part a function of its albedo, that is, the rate at which it reflects the sunlight. Different things reflect at different rates—clouds are different from grass and trees for instance. "Microbial communities of algae and bacteria that make up mats and stromatolites most likely respond to environmental variables that may change the reflective properties of shallow waters and land surfaces." The important thing is that life itself can have a hand in affecting the overall albedo. Life does not sit back and hope that something will happen. "Some microorganisms in the ancient past, and perhaps even now, may have set conditions where organic compounds oxidized to carbon black, which would serve as an ideal light absorber where warmth was needed" (Margulis and Lovelock 1974, 480). Related to this process is the matter of emissivity. Organic objects are not passively black or white. They often vary from place to place, and it is reasonable to suppose that some of this variation may be a function of the planet's need to maintain a fairly stable and bounded temperature. "Ocean surface scum, forest canopy texture, soil particle size, and soil bacteria may affect the emissivity over large regions of the planetary surface" (480).

Next we consider the ways in which organisms can affect the composition of the atmosphere. The microorganisms in the soil are largely responsible for the production of the gases that we find in the atmosphere, and the rates and kinds of production can have major effects on Earth's surface temperature. Gases such as water vapor, carbon dioxide, and ammonia can, thanks to their infrared absorption, reduce the loss of radiation from Earth to space. This is the *greenhouse effect*, whereby the atmosphere keeps the planet a lot warmer than it would otherwise be. Early in our planet's life, the greenhouse effect seems to have been very important; it has become rather less so as the sun sends out more heat. The important point is that the gases vital for the effect are produced by organisms—for instance, ammonia is a major part of organic waste, either in its own right or as uric acid or urea, which in turn break down into ammonia. Levels of ammonia are also controlled by the use that bacteria and fungi make of it, oxidizing it to supply most of the energy they need. There is also the matter of ozone. Its levels are regulated by other gases, such as nitrous oxide. In comes life: "Nitrous oxide is a major biological product; hundreds of megatons a year are released by soil microorganisms" (Margulis and Lovelock 1974, 481). It is true that not much is known about the effects of ozone in the atmosphere, and hence any links between the production of nitrous oxide and climate are at best "tenuous" (482). From the perspective of the Gaia hypothesis, however, we know for sure that the nitrous oxide is not around by chance. There has to be some explanation for its presence, given its fundamental role in the atmosphere and hence in temperature control and moderation. We know that nitrous oxide is produced by microorganisms, but we need more by explanation of its presence than the serendipitous fact that it is produced. If it is needed, what controls the satisfaction of that need? (The solution to this problem is left to the reader.)

Particles suspended in the atmosphere can also affect Earth's temperature. Take clouds, essentially groupings of water droplets. "The area of the earth's surface at present is about 50% covered by clouds and because they are white, they reflect much sunlight back into space. If the proportion of clouds varied, it could affect the Earth's surface temperature quite significantly" (Margulis and Lovelock 1974, 482). There is a feedback system here. More clouds would make things down below cooler; hence there would be less evaporation. Thus, fewer clouds would be formed, which in turn would lead to heating, which would cause more evaporation and thus more clouds. Organisms get involved in this process in several ways. For instance, a layer of steroids or other lipids can inhibit evaporation, and these can be produced by organisms. "Bacteria and algae tend to produce surface scums on ponds and slow running streams. This surface layer is composed of the cells and cell debris" (482). Other particles are also significant. Ammonia gets involved through various reactions with atmospheric gases, causing a kind of aerosol. "Such stratospheric aerosols apparently cause cooling, which explains the drop in tropospheric temperature observed after major volcanic eruptions. Eruptions eject sulfur dioxide into the stratosphere, which oxidizes to sulfuric acid and subsequently forms the ammonium sulfate aerosol" (482).

We must also consider the matter of acidity—or, more precisely, its opposite. Overall Earth is slightly alkaline, which is crucial for such organisms as blue-green algae, which grow very poorly or not at all in acid conditions. Yet this is a bit of a puzzle. One would expect the opposite: the increase of oxygen in Earth's atmosphere suggests that the system should become more acidic. There must be something going on, and Lovelock and Margulis hypothesize that the missing factor is the production of ammonia (which is alkaline). Thanks to organisms, "approximately 2×10^9 tons/yr of ammonia are injected into the atmosphere. Unless the system is under cybernetic control, it is another remarkable coincidence that this is just a little more than sufficient to neutralize all of the acid formed in the oxidation of sulfur and nitrogen compounds" (Margulis and Lovelock 1974, 483). Parallel to this, another important factor in keeping things going here on Earth is how elements are circulated, whether through the air or the waters. It is highly improbable that life would ever arise if there were no atmosphere at all. But the atmosphere and its circulation alone are not enough. The transportation of essential chemicals is necessary. For instance, pure rainwater leaches out nutrient elements that are necessary for life. Fortunately, however, a remedy is at hand. "The release of methylated derivatives of essential elements like iodine, sulfur, and perhaps selenium and phosphorus, we consider a mechanism of transporting these elements back to the land surfaces, a deliberate contrivance of Gaia to circulate them" (483). More than this, the Gaia concept led the writers to look to the sea to find the needed elements, as well as to investigate how the elements might be transported back to land. In the case of phosphorus, for instance, "it may even turn out that airborne spores, birds, insects, and migratory fish are examples of 'biologically released volatiles' of phosphorus" (484).

Finally, we must consider the production and subsequent levels of oxygen. All sorts of mechanisms are involved here, including the fairly obvious. Fire may be a crucial factor, because it uses up oxygen and nitrogen. It does seem that it would be a "last resort" method of regulation, although we cannot rule out the possibility that "regular forest fire[s] may be a local mechanism to remove oxygen and nitrogen from the atmosphere to the biosphere in the form of nitrate" (Margulis and Lovelock 1974, 485). Whatever the details may be, here as elsewhere the conclusion is becoming inescapable. Earth remains fairly stable with respect to temperature, acidity, atmospheric composition, and so forth. It is highly unlikely that this all happens by chance. It is most reasonable to assume that life itself has a hand, a major role in fact, in keeping things balanced for its own benefit. Given the many environmental variables involved, we are led to deny that they "are, by chance alone, precisely those required for the aerobic life that has evolved on earth. It is more reasonable to assume that, at the core at least (namely, the tropical and temperate zones), life has evolved and utilized many mechanisms to keep these variables from exceeding limits that are intolerable to all terrestrial species" (485–86).

Margulis and Lovelock deal rather briefly with how this has all come about. They assure us that there is no claim about a "planetary engineer." Darwinian mechanisms involving natural selection would seem to be the key. Population pressures set up a struggle for existence; this leads to a consequent "selecting" of the fittest, and design-like features, "adaptations," emerge. As life began and persisted, it altered the chemical composition of its habitat. "As soon as metabolism induced physical conditions which were unfavorable, organisms were selected for which grew under such altered conditions. Subsequently new conditions provided opportunities for other variant forms. Those organisms able to maintain or alter conditions to favor their own growth have left more offspring" (Margulis and Lovelock 1974, 486). A balance soon came about. "This evolutionary pattern implies that planetary homeostasis developed early in the history of the planet" (486).

CHANGING THE AUDIENCE

"Homeostatic Tendencies of the Earth's Atmosphere," by the same two authors (Lovelock and Margulis 1974a) was a little more restrained. Although they acknowledged that the term *Gaia* comes from the Greek for "Mother Earth," they defined it rather more neutrally as "the idea that energy is expended by the biota to actively maintain" the optima needed to sustain life. Also written and published at the same time (in a Swedish journal) was an article entitled "Atmospheric Homeostasis by and for the Biosphere: The Gaia Hypothesis" (Lovelock and Margulis 1974b). Although its focus is on the atmosphere exclusively, this article covers much of the same ground. The authors urge an inclusive approach—"The entire ensemble of reactive gases constituting the atmosphere needs to be considered"—which produces answers that differ from those obtained from a purely analytic approach: "Information is made available which is otherwise inaccessible when each gas is considered separately in isolation" (3). Most significantly, the authors state that "this approach applied to the present problem of the anomaly of the chemical distribution of the gases of the atmosphere, offers a strong suggestion that Earth's atmosphere is more than merely anomalous; it appears to be a contrivance specifically constituted for a set of purposes." This leads to "the hypothesis that the total ensemble of living organisms which constitute the biosphere can act as a single entity" regulating environmental factors such as the atmospheric composition, acidity, and perhaps even climate. The Gaia hypothesis emphasizes the "notion of the biosphere as an active adaptive control system able to maintain the Earth in homeostasis." Hence the term refers to "the biosphere and all of those parts of the Earth with which it actively interacts to form [a] hypothetical new entity with properties that could not be predicted from the sum of its parts" (3).

From this article, we gather that the intended audience is changing. At first, Gaia was aimed entirely toward professional scientists. There was a real caginess about coming right out and saying that Earth is an organism. The already-encountered skepticism told. For whatever reason, the hypothesis had to be introduced with stealth. Hence, generally, Lovelock and Margulis played down the kind of language often associated exclusively with organisms, where we try to understand things in terms of what we expect them to do (what we anticipate in the future) rather than what they have done or what caused them (what occurred in the past)—where we talk in terms of *function, purpose,* and *ends* (Ruse 2003). You can ask, for instance, What is the function of the nose or the eye? What do you expect the nose or the eye to do? We do not ask, "What is the function of Mount Everest?" or "What end has the Moon in view?" It is true that Lovelock and Margulis were not strict about this. When talking about the large quantities of nitrous oxide, N_2O, we do get hints of purpose: "In the context of Gaia the production of such large quantities of N_2O requires an explanation of its role in the atmosphere" (Margulis and Lovelock 1974, 482). But in the earlier articles, the language is restrained. In this last article, the language of organism is more explicit: "If we assume the Gaia hypothesis, and regard the atmosphere as a contrivance, then it is reasonable to ask what is the function of its various component gases" (Lovelock and Margulis 1974b, 5). Somewhat self-consciously, the authors admitted to a certain queasiness, but covered themselves: "Outside the Gaia hypothesis such a question would rightly be condemned as circular and illogical, but in its context such questions are no more unreasonable than asking, for example, what is the function of fibrinogen in blood?" (5). With this reassurance, the authors go on to discuss nitrous oxide, explaining the fact that this gas would not be around if the atmosphere were in equilibrium and describing its manufacture by soil microorganisms. "By the Gaia hypothesis it must have an important atmospheric purpose; could this be concerned with the regulation of the position or density of the ozone layer? In the summary there is overwhelming evidence that the atmosphere, apart from its content of noble gases, is a biological product" (6). They add, "It [nitrous oxide] may also be a biological contrivance; not living but as essential a part of the biosphere as is the shell to a snail or the fur to a mink" (6). Whatever the full role of nitrous oxide, again it is appropriate to invoke the notion of homeostasis. There has to be more involved than just getting to chemical equilibrium. "If life actively cycles the gases, then we ask how could such a system be stable in the long run without homeostasis?" (9). And, as they do again and again, Lovelock and Margulis argue that it is by focusing on the state of homeostasis that we find the definition of life. As three-sidedness is the essence of triangularity, so homeostasis is the essence of life.

We are seeing a shift here, especially on the part of Lovelock, who was, after all, the original author of the Gaia hypothesis. He wanted to get his idea out in the open domain, and if the professional scientific community would not entertain it, he would look to others who might. In 1975, Lovelock addressed the general public in an article announcing Gaia in *The New Scientist*, the popular British science magazine. The language here was a lot less cagy. We are moving from the world having life to the world being life. An introduction (presumably added by the magazine editors) asked whether "the Earth's living matter, air, oceans and land surface form part of a giant system which could be seen as a single organism." The answer is unambiguous: "Living matter, the air, the oceans, the land surface" are parts of a giant system that keeps things functioning at an optimum, thus exhibiting "the behavior of a single organism, even a living creature" (Lovelock and Epton 1975, 304). Then, responding to requests sparked by that article— twenty-one publishers wanted a book—Lovelock published a full-length work in 1979, *Gaia: A New Look at Life on Earth*, intended for the general public. In this popular writing, Lovelock was far less restrained. Starting with the exploration for life on other planets, he wrote, "This book is also about a search for life, and the quest for Gaia is an attempt to find the largest living creature on Earth" and went on to say that "if Gaia does exist, then we may find ourselves and all other living beings to be parts and partners of a vast being who in her entirety has the power to maintain our planet as a fit and comfortable habitat for life" (Lovelock 1979, 1). He devotes much space to recapitulation (in a significantly more reader-friendly fashion) of earlier discussions. He introduced, however, an interesting new line of argument about the salt in the sea—or, more precisely, about the salt that is not in the sea. The salinity of the sea is about 3.5%. If you calculate how much salt is being dumped into the sea yearly by runoffs from the land (via rivers), it is clear that this percentage should be a lot higher. In fact, in a mere eighty million years, starting from scratch, salinity could reach present levels—not very long when you consider how long the oceans have been around (about four billion years). Where is all of the salt going? And why? Of one thing we can be certain: salt levels can never have been very much higher than they are now. Apart from a few odd organisms (e.g., the brine shrimp, with its stupendously tough and impenetrable shell), most organisms start dying out at 4%, and for them 6% is the absolute tops. Inorganic processes don't seem up to the job of maintaining the current level: "It is surely time to ask ourselves whether the presence of the living matter with which the seas abound could have modified the course of events and may still be acting to solve this difficult problem" (Lovelock 1979, 87).

Lovelock's suggestion focuses on the microorganisms in the sea, and specifically on diatoms, algae with skeletal walls made of silica. Hypothesizing from the fact that dead, landlocked salt lakes have high proportions of silica, Lovelock suggested that diatoms flourish on the surface of the seas and sink when they die, carrying down with them micro-quantities of salt. Over the years, these deposits build up but are locked away on or beneath the oceans' floors. This gets rid of the salt, but there is more to the situation than that, because salt levels must be kept fairly constant to sustain the life that has evolved in the seas.

> This biological process for the use and disposal of silica can be seen as an efficient mechanism for controlling its level in the sea. If, for example, increasing amounts of silica were being washed into the sea from the rivers, the diatom population would expand (provided that sufficient nitrate and sulfate nutrients were also in good supply) and reduce the dissolved silica level. If this level fell below normal requirements, the diatom population would contract until the silica content of the surface waters had built up again, and this is well known to occur. (Lovelock 1979, 88–90)

Referring to the dying diatoms, Lovelock added, "This deluge of dead organisms is not so much a funeral procession as a conveyor belt constructed by Gaia to convey parts from the construction zone at surface levels to the storage regions below the seas and continents" (90).

Giving his imagination further free reign, Lovelock then went on to suggest that life also constructs lagoons in which sea water can be trapped and evaporated. The salt from the main bodies of water is removed this way. How does the construction occur? Coral reefs are one possibility. "Is it possible that the Great Barrier Reef, off the north-east coast of Australia, is the partly finished project for an evaporation lagoon?" (Lovelock 1979, 91) And what about the building power of stromatolites (huge mats of blue-green algae)? They could be built up and act as the containers for lagoons also. It is even possible that life manipulates the inorganic world. Volcanic activity and continental drift could be turned to nature's ends. Perhaps the falling silica puts more and more pressure on the surface of the earth, and at the same time the silica forms a kind of insulating blanket that prevents heat from escaping. Pressure and heat down below build up and up until there is an explosion and things are thrown up from below by volcanic activity. "Volcanic islands could be formed in this manner, perhaps lagoons as well" (92). Lovelock added, "I am not, of course, suggesting that all or even most volcanoes are caused by biological activity; but that we should consider the possibility that the tendency towards eruptions is exploited by the biota for their collective needs."

We are writing in the popular domain, so we are bound to ask, "Where do humans fit into any of this?" Unfortunately—although hardly atypically in works such as these—it turns out that we are terrible threats to the well-being of Gaia. In some respects, Lovelock was almost blasé about human actions. "The very concept of pollution is anthropomorphic and it may even be irrelevant in the Gaian context" (Lovelock 1979, 110). Indeed, he was getting into hot water around this time because, having—thanks to his inventions—been the person to discover the widespread existence of chlorofluorocarbons (CFCs) in the atmosphere, which led to the general panic about their contribution to the shrinking ozone layer above Earth, Lovelock paradoxically dismissed their significance. Given the use of CFCs in aerosols and refrigerators, critics charged that Lovelock's judgment was warped because of his links to industry. More likely he was seduced by the presumed power of Gaia. But this did not mean that Lovelock thought human actions irrelevant or incapable of harm. We do all sorts of things to upset the balance—for instance, clearing forest land and scrub by burning. In addition, our activities release all manner of pollutants into the air that affect its composition. And other seemingly innocuous activities—actual or proposed—are even worse. It has been suggested, for instance, that we might grow vast amounts of kelp in the seas. Whatever the benefits—kelp can be refined to produce many useful chemical compounds—we could thereby reduce the iodine levels in the seas, with disastrous secondary effects. Many other necessary minerals, such as sulfur and selenium, could be removed, to the detriment of land animals. The dangers come not so much from pollutants, because in fact many of these are already produced naturally. It is in ignorance of Gaia that the real threat lies. "There is only one Pollution . . . People" (114).

Can nothing positive be said about us? What of the obvious suggestion that we humans collectively form the thinking apparatus, the intelligence, of Gaia itself? Not entirely enthusiastically, Lovelock conceded that there might be something to this. Having talked about how we can plan for the future—taking warm clothes for a winter trip to New Zealand, for example—he wrote, "So far as is known, we are the only creatures on this planet with the capacity to gather and store information and use it in this complex way. If we are part of Gaia it becomes interesting to ask: 'To what extent is our collective intelligence also a part of Gaia? Do we as a species constitute a Gaian nervous system and a brain which can consciously anticipate environmental changes?'" (Lovelock 1979, 138–39) He went on to add, "Whether we like it or not, we are already beginning to function in this way." Suppose a large object were heading toward Earth. With our modern technology, it is at least possible that we might deflect it. Using rockets and hydrogen bombs, we might be able to change a certain hit into a near miss. Perhaps more excitingly, we have brought a level of self-awareness to Gaia. With this could come a sense of integration into the whole. "It may be that the destiny of mankind is to become tamed, so that the fierce, destructive, and greedy forces of tribalism and nationalism are fused into a compulsive urge to belong to the commonwealth of all creatures which constitutes Gaia" (140).

Lovelock was not joking. He seemed tremendously impressed by the size of the brain of the whale, many times larger than the human brain. Possibly in the future we shall be able to harness whale brains, in some sense integrating them with human brains— one presumes in function rather than anatomy—and get something altogether bigger and better. "Perhaps one day the children that we shall share with Gaia will peacefully co-operate with the great mammals of the ocean and use whale power to travel faster and faster in the mind, as horse power once carried us over the ground" (Lovelock 1979, 142). There is simply no end to the prospects opened up by the Gaia hypothesis.

2

THE PARADOX

Whether you accept the Gaia hypothesis or not, Lovelock and Margulis were clearly addressing some tremendously interesting and important issues. Although the casual reader might have missed the full import—the very striking figure 1, which redraws the original figures, is larger and clearer than the originals—the message is there for all who would read. Earth is not heating up as one might expect. More than that, Earth is maintaining the kinds of conditions that support life, from the most primitive forms down to humans. The sorts of solutions Lovelock and Margulis were offering were not *a priori* stupid or subject to instant scorn and dismissal (why not some feedback temperature control involving albedo?), and they had each—especially by the end of the decade (around 1980)—earned the right and status to be taken seriously. And yet—and here is the first of the reasons why the Gaia story is worth reconsidering three decades later—the expected reactions were very different from the actual reactions. Let us start with the scientists.

THE BIOLOGISTS TAKE NOTICE

We have seen already that, from the beginning, the professional scientific community was at best amused and generally deeply skeptical. Gaia was something for light relief at the end of a hard day's work. No need to take it seriously. With the publication of his more popular book, it was clear that Lovelock was not going to shut up and go away. Something had to be done, and at this point the heavyweights—people who had reason to feel qualified to speak on these issues—swung into action.

With hindsight, it was almost predictable that the English biologist Richard Dawkins would devote critical time to Gaia. His best-selling book *The Selfish Gene*, published in 1976, had rocketed him from obscurity to a degree of fame that still persists. Dawkins was the public face of science, the scourge of the sloppy and inadequate. Given that Gaia stresses the interconnectedness of things, one might have expected Dawkins to react favorably. A major message of *The Selfish Gene* is the extent to which organisms are not isolated units, but parts of wholes, functioning together. In the words of the great metaphysical poet John Donne, "No man is an island," and that applies throughout the living world, especially the animal world. Moreover, like Lovelock and Margulis, Dawkins believed this was not a matter of chance. It was the result of the Darwinian process of natural selection: population pressures mean that there is an ongoing struggle for existence and reproduction, and this leads to differential reproduction (natural selection), resulting in the development of adaptations that aid them in the struggle. For Dawkins, as for Lovelock and Margulis, the adaptations were precisely those that led to cooperation that resulted in the greater good of all.

In his second book, *The Extended Phenotype*, published in 1982, Dawkins continued to stress the interrelatedness of life, given natural selection, but argued (apparently even more in agreement with the Gaia proponents) that the inorganic world could also be involved in the picture. For instance, the dams built by beavers to create ponds in which to build their lodges can be considered as adaptations, just as hands or eyes are. Moreover, this can involve pretty extensive swathes of territory. "A beaver that lives by a stream quickly exhausts the supply of food trees living along the stream bank within reasonable distance. By building a dam across the stream the beaver quickly creates a large shoreline which is available for safe and easy foraging without the beaver having to make long and difficult journeys overland" (Dawkins 1982, 200).

Dawkins added that a whole squad of beavers is often involved, working together to dam and to build.

With all of this interrelatedness among organisms and between organisms and the environment—relatedness that Dawkins hypothesized could even cross continental boundaries—we seem well on the way to Gaia. Yet Dawkins wanted nothing to do with the Lovelock-Margulis hypothesis. Absolutely nothing. The reason is simple—and (as we shall see later) extremely important to our story. The Gaia hypothesis as it was being presented took the good of the whole as the fundamental driving force, the reason why everything else falls into place. According to the Gaia hypothesis, things often do not exist or behave for their own benefit, but for the benefit of the whole. Thus, for instance, plants produce oxygen because it is necessary for animal life. And, concerning methane, "One obvious function is to maintain the integrity of the anaerobic zones of its origin" (Dawkins 1982, 235, quoting Lovelock 1979, 73). And concerning nitrous oxide, "We may be sure that the efficient biosphere is unlikely to squander the energy required in making this odd gas unless it has some useful function. Two possible uses come to mind" (quoting 74). Finally, with regard to ammonia: "As with methane, the biosphere uses a great deal of energy in producing ammonia, which is now entirely of biological origin. Its function is almost certainly to control the acidity of the environment" (quoting 77).

For Dawkins, such thinking gets the workings of natural selection precisely backwards. In his opinion natural selection never acts primarily for the good of the group, as is clear from the title of his first book. It always acts first for the good of the individual; benefits for the whole arise as fortunate side effects. It might well pay us to be "altruistic," as biologists put it, but only if we benefit more by such actions than otherwise. For instance, I contribute to a group insurance scheme, even though I may never benefit from it. But I am covered in case I do fall ill or become incapacitated. I am not in the scheme for your well-being but for my own. We have a group benefit only because of individual "selfish" interests. Translating this into the language of the units of heredity—which evolutionists have been wont to do ever since Darwinian selection was melded with Mendelian (now molecular) genetics in the 1930s—although the effects are altruistic, the genes are selfish. "Individual selection" always tops "group selection." It is here that Gaia comes a cropper. "The fatal flaw in Lovelock's hypothesis would instantly have occurred to him if he had wondered about the level of natural selection process which would be required in order to produce the earth's supposed adaptations. Homeostatic adaptations in individual bodies evolve because individuals with improved homeostatic apparatus pass on their genes more effectively than individuals with inferior homeostatic apparatuses" (Dawkins 1982, 235–36).

In any case, in Dawkins's opinion, there is something wrong with the emphasis that Gaia puts on homeostatic self-regulation as the defining characteristic of living things. Dawkins was happy to accept the connection between homeostasis and life, but he wanted to counter the belief that homeostasis alone is enough to deem something as living. Organisms have to be the end product of natural selection. And once you recognize this, you see how flawed the Gaia hypothesis has to be. "For the analogy [of the Earth as an organism] to apply strictly, there would have to have been a set of rival Gaias, presumably on different planets. Biospheres which did not develop efficient homeostatic regulation of their planetary atmospheres tended to go extinct. The Universe would have to be full of dead planets whose homeostatic regulation systems had failed, with, dotted around, a handful of successful, well-regulated planets of which the Earth is one (Dawkins 1982, 236). Homeostasis, a feature of organisms, must, like all features, be the consequence of selection brought on by reproduction, rather than the initial spur of such reproduction. Dawkins added that even then we are left with the problem of how planets reproduce. And you cannot just say that natural selection works on our planet to produce Gaia, because that plunges you right back into the unacceptable group-selection position. A plant that decided not to produce oxygen for animals (just for the sake of animals) would be ahead of those that do—in the short run it would be serving its own interests in the struggle for existence and reproduction—even though down the road everyone might suffer when animals went extinct.

Similar critiques came from others, including the North American biologist W. Ford Doolittle. He was clearly bothered by the aura of intention or planning that hung over the Gaia hypothesis, making semi-joking reference to children's books about a near-namesake, Dr. Dolittle, especially to one story that involved a Council of Life on the moon that prevented fighting and promoted harmony. In the results of the Council's deliberations, we clearly have ends (if you like, intended ends) in some way influencing earlier events, a kind of planning. How can we suppose something like this for Earth? He saw that group selection was going to pick up some of the slack here, promoting the kind of unity and harmony that consciousness or design might be expected to do. And, like Dawkins, he could not buy this, arguing that there is no way in which Earth or its components could have the forethought to do things that would have good consequences generations down the road. "The construction of an evaporation lagoon for sequestration of sea salt may benefit the biosphere as a whole, in the very long run, but what in particular does it do for the organisms who construct it, especially in the short run?" (Doolittle 1981, 61). And in any case, why should one assume that the way the world keeps things in balance is the best or optimal way? "The global conflagration expected if oxygen levels exceed 25% would be disastrous to most higher forms of life. But it would produce a large amount of carbon dioxide and consume a lot of oxygen, and it is carbon dioxide which is the life giving substrate for the methanogens, and it is oxygen which they must scrupulously avoid (because it is toxic to them). Would methanogens not in fact benefit, at least for thousands of years, from such a disaster?" (61)

Doolittle's judgment was that although Jim Lovelock's "engaging little book" gives one "a warm comforting feeling about Nature and man's place in it," it is based on a view of natural selection "which is unquestionably false." Even worse, it is potentially dangerous, because it gives the illusion that if things go wrong,

Earth will fix itself, whereas in truth there is absolutely no guarantee. Things work by chance, not design. Any sense of design comes about through illusion—if things did not work as they do, we would not be here. That is all. "Only a world which behaved as if Gaia did exist is observable, because only such a world can produce observers" (Doolittle 1981, 62). To assume that there is more could lead us into the trap of behaving as if there will inevitably be a tomorrow, when in truth tomorrow may never come. Dawkins, incidentally, echoes this concern. He argues that Gaia commits the fallacy of the "BBC Theorem," meaning that it assumes (as do nature programs shown on the BBC) that the world is a harmonious whole, threatened only by humans and their desperate striving for endless progress. Supposedly all works together happily; in the words of the Reverend Reginald Heber, "Every prospect pleases, and only man is vile." In Dawkins's opinion, there is simply no justification for this kind of thinking—optimistic about the "natural" way of things, if pessimistic about the effects of humans on the whole.

Gaia was under attack. I do not want to downplay or trivialize these criticisms. Dawkins particularly, then as now, was controversial. But he was expressing a line of thought that had gained great credence in the two decades before he started writing, and it had serious support. At the theoretical level, many biologists interested in social behavior were pointing out that group selection is open to the seemingly fatal objection of cheating (Ruse 1979, 2006). Suppose you have two otherwise similar organisms, one working for itself and the other working for the group. The problem is that the organism working for itself is going to get all of the benefits from its own labors and some of the benefits from the labors of the other. The organism working for the group will get nothing from the selfish organism and diminished benefits from its own labors, which are at least in part directed to others. Biologically speaking, therefore, the selfish organism is going to be at an advantage in the struggle for survival and reproduction, and in a generation or two will have quite eliminated the other. At the empirical level, study after study was backing up this kind of thinking. From the social insects through the lower animals, to the mammals and then on to the great apes, thinking in terms of individual rather than group selection was paying great dividends. "Gaia got such a bleak reception from them [evolutionary biologists] in the seventies and eighties . . . because they were in a phase of trying to get rid of all group selectionist-type woolly thinking from their subject, and Gaia came along at just the wrong moment as the most extreme form of group selectionism imaginable. So in a way it was just bad timing" (TL).

Yet one senses that there was something more. The disagreement over Gaia does not seem to have been just a matter of pure science. For a start, although there were strong arguments against the original Gaia formulation, the debate was not all one-sided. There were then (and over the years the numbers have grown) people who were ready to argue on behalf of group selection (Wilson 1980; Sober and Wilson 1997). They thought that in important cases the strong benefits of behavior directed selflessly to others could overwhelm the immediate costs of such behavior. The maintenance of sexuality might be one such case. Working in the tradition mentioned above of conceptualizing evolution in terms of the genes, think therefore of natural selection not so much in terms of individual organisms reproducing or not, but of their units of heredity (their genes) being transmitted or not from one generation to the next. From the female's perspective, sex seems a bad bargain. If she reproduced asexually, then each child would be carrying a complete set of her genes. By reproducing sexually, she is giving up half her reproductive output to males, a seemingly silly thing to do, especially since most males do not compensate with shared effort in child-rearing. But of course the big advantage to sexual reproduction is that good new variations (mutations) that improve an organism's chances in the struggle for reproduction can be passed quickly through the population, and all can benefit. Perhaps, therefore, the virtues of the sexual process are so great that the individual cost is balanced by the group benefit. Supporting this suggestion is the empirical finding that among species with both asexual and sexual strains, the former tend not to persist as long (Maynard Smith 1978).

THE VITRIOLIC DEPTH OF REACTION

The debate was thus not quite so one-sided as one might think. In any case, scientific criticisms are one thing, and sheer nastiness is quite another. Dawkins is always capable of being, shall we say, "robust," but others joined in like fashion. As Lovelock says truly of reactions to Gaia, "The biologists hated it right from the beginning. They loathed it" (JL). And he is not kidding. Despite John Maynard Smith's suggestion that sex may be one case in which group selection can have a beneficial effect, his reaction was as violent as it was typical: "Gaia is just an evil religion" (BL). Nor were others much more complimentary. Like Richard Dawkins in Britain, the Harvard paleontologist Stephen Jay Gould was the public face of science in the United States. As it happens, he was another who was less than fully sympathetic to wholesale individual selection. For a start, he was proposing some kind of "species selection," in which a form of differential reproduction takes place between whole groups rather than within groups. This is not interplanetary biosphere selection of the kind that earned Dawkins's scorn, but it is certainly selection at a much higher level than we find in *The Selfish Gene*. None of this stopped Gould from being as dismissive of the Gaia hypothesis as those already mentioned. In one of the brilliant essays he penned for the journal *Natural History*, Gould wrote that he liked "to apply a somewhat cynical rule of thumb in judging arguments about nature that also have overt social implications: when such claims imbue nature with just those properties that make us feel good or fuel our prejudices, be doubly suspicious" (Gould 1987, 21). He continued, "I am especially wary of arguments that find kindness, mutuality, synergism, harmony—the very elements we strive mightily, and so often unsuccessfully, to put into our own lives—intrinsically in nature," explaining that "Gaia strikes me as a metaphor, not a mechanism."

Somewhat condescendingly, he concluded, "Metaphors can be liberating and enlightening, but new scientific theories must supply new statements about causality. Gaia, to me, only seems to reformulate, in different terms, the basic conclusions long achieved by the classically reductionist arguments of biogeochemical cycling theory" (21).

Indifference. Hostility. Rejection. Sneering. Outright condemnation. Papers blocked from publication and derision all around. Robert May, future president of the Royal Society, member of the British House of Lords, and holder of the Order of Merit, called Lovelock a "holy fool" (JL). Although in print he was more negative than brutal, Paul Ehrlich, American biologist and prophet of doom about population growth, was really rough in person. He "hated Gaia" (JL). It wasn't just wrong. He saw Lovelock as "rather more radical and dangerous." John Postgate, leading British microbiologist, Fellow of the Royal Society, and colleague of John Maynard Smith, exclaimed, "Gaia—the Great Earth Mother! The planetary organism! Am I the only biologist to suffer a nasty twitch, a feeling of unreality, when the media invite me yet again to take it seriously?" He continued that Gaia "has metamorphosed, in Lovelock's writings and those of others, first into a hypothesis, later into a theory, then into something terribly like a cult." In his judgment, it was "pseudoscientific mythmaking." And so came the warning. "When Lovelock introduced it in 1972, Gaia was an amusing, fanciful name for a familiar concept; today he would have it be a theory, one which tells us that the Earth is a living organism. Will tomorrow bring hordes of militant Gaia activists enforcing some pseudoscientific idiocy on the community, crying 'There is no God but Gaia and Lovelock is her prophet'? All too easily" (Postgate 1988, 80).

Obviously, we do not yet have all of the story or even all of the puzzle; but one thing is certain. There must be more to things than Gould suggests. The problem cannot be metaphor, or at least not metaphor as such. After all, we are dealing with critics who venerate a theory where the main mechanism is natural selection, brought on by a struggle for existence. Charles Darwin himself was the first to explain carefully that there is certainly no one out there doing any selecting. That is the whole point of Darwin's naturalistic theory.

[It has been] objected that the term *selection* implies conscious choice in the animals which become modified; and it has even been urged that as plants have no volition, natural selection is not applicable to them! In the literal sense of the word, no doubt, natural selection is a misnomer; but who ever objected to chemists speaking of the elective affinities of the various elements?—and yet an acid cannot strictly be said to elect the base with which it will in preference combine. It has been said that I speak of natural selection as an active power or Deity; but who objects to an author speaking of the attraction of gravity as ruling the movements of the planets? Every one knows what is meant and is implied by such metaphorical expressions; and they are almost necessary for brevity. (Darwin 1861, 65)

Metaphors persist in evolutionary thinking. In the twentieth century, probably the most fertile idea floated in evolutionary circles was Sewell Wright's (1932) notion of an adaptive landscape, a picture that sees genes metaphorically atop an undulating surface, with hills and valleys that—through a combination of drift and selection—the genes move onto or into. Nobel Prizes were won after the cracking of the code of the DNA. And moving to the present, it is almost superfluous to mention that genes are not literally selfish or that no one was as gifted as Stephen Jay Gould when it came to the appropriate turn of phrase, a practice of which he approved wholeheartedly. "Our mind works largely by metaphor and comparison, not always (or often) by relentless logic" (Gould 1991, 264).

The talk of metaphor gives us a clue about the other side to the reactions to Gaia. Metaphors make things easier to understand (Ruse 2005). The adaptive landscape picture was powerful—even though, as many have noted, it has grave conceptual flaws—because people could understand it (Ruse 2004). Particularly in an age when many biologists lacked fully developed mathematical skills, metaphor provided a compelling insight into the way that evolution works. Likewise, a major factor in Dawkins's success with *The Selfish Gene* was that, like it or hate it, when he described the units of heredity as "selfish"—even though literally such talk was clearly false—he brought home in a resoundingly frank manner how he and his fellows were thinking about the actions of natural selection: it works for the individual, not the group. It was the same with Gaia. Talk about homeostasis and the self-regulation of gases— topics vaguely remembered from high-school chemistry—can be pretty tough going. What's the difference between acidity and alkalinity, and who cares? Albedo? Move on please! But talk about Earth as an organism—living, breathing, weeping, sweating, farting (don't laugh, it's coming up), and possibly dying—grabs the imagination. Anyone can understand that. And here we begin to get to the heart of the matter. It was not just that the professional scientists hated Gaia; it was also that the general public loved it!

THE PUBLIC DOMAIN

We already have a sense of how this whole debate or controversy had edged from the professional to the public domain. Lovelock had begun it, moving from technical journals to a science magazine and then to a general-audience book. And we have seen how, if implicitly, the critics were responding in the public domain. *The Extended Phenotype* was perhaps more directed to Dawkins's fellow biologists than was *The Selfish Gene*, but still it was a book that the general public could read and appreciate. Gould, of course, published in *Natural History* for a broad audience. And Doolittle is perhaps the most revealing of all. His review appeared in *Co-Evolution Quarterly*—published by the owners of the *Whole Earth Catalog*—sandwiched between a review of a book called *The Divine Woman* ("A true wish book. A study of shamans and rain goddesses.") and an enthusiastic endorsement of homeopathy (which uses minute amounts of poisons to combat disease).

If the hysterical opposition to Gaia by professional scientists is the first part of our puzzle, the enthusiasm of the general public is the second part. The negative reaction by those who might have been most likely to respect and appreciate Lovelock and Margulis was balanced by a public appreciation and embracing of Gaia. In his autobiography, written a couple of decades later, Lovelock freely admitted that the publication of *Gaia: A New Look at Life on Earth* "completely changed my life and the fall of mail through my letterbox increased from a gentle patter to a downpour, and has remained high ever since" (Lovelock 2000, 264). Perhaps showing naïveté, or perhaps with the benefit of hindsight, Lovelock opined, "I never intended the book as a science text for specialists, but I did expect them to read it." No such luck. But there was compensation. "To my astonishment, the main interest in Gaia came from the general public, from philosophers and from the religious. Only a third of the letters were from scientists." Admittedly, to a certain extent, Lovelock and Margulis were preaching to the converted. He was not the only one thinking about "Mother Earth." There was a rock band by that name, and there was (and still is) the magazine *Mother Earth News*, started in 1970. Packed with all sorts of useful information about building eco-friendly houses, alternative fuels, farming and gardening, and pickling and canning, with a large readership (perhaps largely composed of urban dwellers living these delights vicariously), its philosophy was totally in tune with the ideas of Lovelock and Margulis. In the words of one of the cofounders, "I think that we live in an unbelievably marvelous Garden of Eden. Surrounded by miraculous life forms almost without number. Kept alive by a mysteriously interwoven, self-replenishing support system that, with all our scientific 'breakthroughs,' we still do not understand" (Shuttleworth 1975, 10).

Even if one explains the enthusiasm for Gaia in terms of already existing fertile soil, the question remains, Why was the soil fertile? Why was there so much enthusiasm for Gaia-type thinking in the general domain, given the repulsion by professionals? And why, after Lovelock published, did the enthusiasm grow, regardless of the criticisms of the professionals? The Gaia hypothesis—the very

name itself—caught fire and became somewhat of an overnight
sensation. Earth as an organism was just the vision, just the metaphor, for which many individuals and groups were searching. In-
stitutes sprang up; one, for example, was associated with the Cathedral of St. John the Divine in New York City. On Mother's Day
(May 10) in 1981, the Cathedral hosted a celebration of Mother
Earth, featuring *Missa Gaia*, an ecological and ecumenical mass,
with music composed by Paul Winter. "The Alaskan tundra wolf,
whose voice this Kyrie was based on, sings the same four-note howl
seven times in an interval known as the tritone—the sax, tenor
solo voices and chorus answering." Closely related to these people
was the Lindisfarne Association, a group organized for the "study
and realization of a new planetary culture" that has included an-
thropologist Gregory Bateson, economist E. F. Schumacher, and
religious scholar Elaine Pagels. A publishing house was founded
under the name Gaia, and it put out the highly successful *Gaia:
An Atlas of Planet Management*. Others got on the bandwagon in
various ways. In its prime, the Commonwealth Institute in London (now, alas, but a memory) promoted cultural education in an
effort to unite the various peoples of what was formerly the British
Empire. In the hope of moving away from the dominance of Western ideology and practice, it sponsored a winter festival featuring a
"Gaia Song," designed to banish the forces of darkness and point
to a happier future.

Gaia is the one who gives us birth.
She's the air, she's the sea, she's Mother Earth.
She's the creatures that crawl and swim and fly.
She's the growing grass, she's you and I.
(Joseph 1990, 66)

A quick survey of online booksellers suggests that, even thirty
years later, the popular interest in Gaia remains high. Lovelock has
a recent book called *The Vanishing Face of Gaia: A Final Warning*
(2009). A year or two earlier, Lynn Margulis was writing an enthusiastic introduction ("inclusive, tolerant and enlightened") to a

book on Gaia called *Animate Earth: Science, Intuition, and Gaia* (2006) by Stephan Harding. And then there are *The Gaia Project: 2012; The Earth's Coming Great Changes* (2007), by Hwee-Yong Jang ("the Earth is already undergoing a purification process"); and *Earthy Realism: The Meaning of Gaia* (2007), an edited volume including "Can Gaia Forgive Us," by Anne Primavesi ("We have hurt, victimized, damaged the more-than-human community of life on Earth. If so, Gaia's purposes and ours will be served by a repentance that acknowledges the wrong done and effects [*sic*] change in our lives that will heal relationships with the members of that community" [72]). At a somewhat more practical level, there is *Gaia's Garden, Second Edition: A Guide to Home-Scale Permaculture* (2009), by Toby Hemenway ("working with nature, not against her"). More broadly, we find the Gaia Napa Valley Hotel and Spa ("a California eco-friendly property"); Gaia Herbs ("the purity, integrity, and potency of our herbs are the heart of the Gaia difference"); and Gaia Online (an "online hangout, incorporating social networking, forums, gaming and a virtual world").

The Gaia hypothesis attracted and continues to attract many who are unconventional, to say the least. Austrian-born physicist Fritjof Capra followed up his 1975 book *The Tao of Physics* ("Science does not need mysticism and mysticism does not need science, but man needs both") with *The Turning Point* (1982). Bemoaning the inadequacies of modern science, Capra gives warm treatment to Lovelock's science and conclusions. "The earth, then, is a living system; it functions not just like an organism but actually seems to be an organism—Gaia, a living planetary being" (285). This earned a friendly review (in the *New Scientist*, later used as a blurb on the paperback) from Lovelock. "This splendid and thoughtful book is an essential guide for anyone inquiring about the place of science and metascience in our contemporary culture" (Lovelock 1982). Such an encomium no doubt compensated somewhat for physicist Jeremy Bernstein's (1982) comments on Capra's earlier book. Agreeing that science does not need mysticism and mysticism does not need science, he wrote: "What no one needs, in my opinion, is this superficial and profoundly misleading book."

Rupert Sheldrake is even more beyond the pale. He diverted the makings of a brilliant scientific career—double first (first class in two subjects) from Cambridge, research fellow of the Royal Society—into the writing of *A New Science of Life: The Hypothesis of Morphic Resonance* (1981). The editor of *Nature* wrote that his book was "an exercise in pseudo science" and asked if this was "a book for burning?" A follow-up book, *The Rebirth of Nature: The Greening of Science and God*, has Gaian fingerprints on virtually every page: "Mother Nature is reasserting herself whether we like it or not. In particular, the acknowledgement that our planet is a living organism, Gaia, Mother Earth, strikes a responsive chord in millions of people; it reconnects us both with our personal, intuitive experience of nature and with the traditional understanding of nature as alive" (Sheldrake 1991, 10).

It would be unfair to suggest that the only Gaian enthusiasts are vegans wearing uncomfortable sandals. If there is one thing that characterizes the philosopher Mary Midgley, editor of the above-mentioned *Earthy Realism*, other than a fiercely independent intelligence, it is a down-to-earth, brisk common sense. She is devoted to puncturing the pretensions of the high and mighty, to showing how the most confident of assertions are too often based on foundations of very dubious worth. Of Richard Dawkins on selfish genes she used the terms "confusing," "weakness," "uncritical," "unworkable," "over-simplified," "bankrupt," "speculations"— and that was before she got to page two (Midgley 1979, 439). She has long been sympathetic to and now is increasingly outspoken in her passion for Gaia. Even before Lovelock published his first book, she was arguing for a more inclusive view of humans and their environment. And her passion is undiminished. "Man needs to form part of a whole much greater than himself, one in which other members excel him in innumerable ways. He is adapted to live in one. Without it, he feels imprisoned; the lid of the ego presses down on him" (Midgley 2005b, 374). When she first grasped the Gaia hypothesis, Midgley's joy knew no bounds. She saw the virtues of the science of Earth, but she wanted also to put this in a broader context, to argue that we can transcend the selfish view that each individual works only to maximize personal ends. "The metaphysical idea that only individuals are real entities is still present in this picture and it is always misleading. *Wholes and parts are equally real*" (365; emphasis in original). More recently she has said that, "like babies, we are tiny, vulnerable, dependent organisms, owing our lives to a tremendous whole." Mary Midgley, of course, is not representative of everyone, but she clearly is not unique. Her edited volume *Earthy Realism* includes contributions by philosophers, climatologists, biochemists, and others. There were and still are many people who just love Gaia.

QUESTIONS

I will say more later about others who responded favorably to Gaia, but for now we have enough to articulate our puzzle. Lovelock and Margulis were serious, sensible scientists for whom respect was growing—Lovelock already was a man of substance, and Margulis was rising rapidly to major heights. Why then was the scientific community so indifferent (and then hostile) toward the Gaia hypothesis? Why did Lovelock feel that he had to go the route of popular science? In his autobiography he says, "Science affects our lives and that of the Earth so much that it would be monstrous for it to retreat to a world of jargon accessible only to the denizens of cosy ivory towers. I wrote the book as if it were a long letter about Gaia to a lively intelligent woman" (Lovelock 2000, 264). That may be true, but his technical papers could hardly be read by any untutored being, however intelligent, of either sex. Why did he feel (with some considerable justification) that he had to switch audiences to get attention? And why, when he had switched audiences, did he get attention?

In part, we know already the reason for this. In the 1970s people had become interested in ecology and the environment, in the status of Planet Earth, and in metaphors that started to pull things together in integrated wholes. Independently of Lovelock and Margulis, many people were grasping for Gaia-like hypotheses. When scientists of stature started to endorse them, they were delighted. There was already something "in the air" with which Lovelock's thinking resonated. And note that when scientists, biologists in particular, did wake up and realize that they had to go on the offensive, it was mainly in this popular domain that they felt they had to respond. Or, to be accurate, mainly in the popular domain that they did respond. Obviously, Doolittle was writing precisely because the Whole Earth people cared about the ideas and wanted discussion of them. So our question is, Why was there something in the air, or (to revert to the earlier metaphor) why was the soil so fertile?

This is interesting. Gaia was produced by professional scientists with status. This had to give the hypothesis its own derived status at some level. Sociologists of science would point out that the "Matthew Principle" kicks in: "To him that hath shall be given, and from him that hath not shall be taken, even that which he hath." You can ignore a grad student at a minor university. You do not ignore Fellows of the Royal Society. Even if you don't like the ideas, you give them some respect. Yet it is clear that there was something about Gaia that did not appeal, or worse. It was not that it was published and then knocked down. It really did not get going—an ex-husband had to take pity on one of the proposers. "No matter how I tried to persuade scientists that they should take Gaia seriously, I rarely succeeded." And things did not improve. "Gaia was treated more as science fiction than science, and it became almost impossible to publish a paper with Gaia in the title or even in the text, unless it was to denounce it" (Lovelock 2000, 267). Yet, Gaia got a warm reception in the popular realm, a realm flavored by the distinct culture of the 1960s and 1970s. But this fact intensifies the paradox. You might think that as a Berkeley graduate, Margulis would be readily explicable. However, she graduated before the campus became the symbol for every radical idea America ever embraced. While there, when she was not working, she was looking after small children. Lovelock is even more puzzling. His idea came first out of a strand of the 1960s against which much of the popular culture—the part that would embrace Gaia—was violently opposed. Lovelock was into space science, not beads and long hair. His funding came from the military and from industry. He was a man whose livelihood depended on making machines that work, whether for helping people or killing them. Not that this helped on the other side of the debate. For many senior scientists, this made things even worse. There was something almost treasonable about Lovelock's thinking. "Gaia was not only wrong; it was dangerous. They saw it as a topic like astrology that masqueraded as a science but was nonsense" (267).

There is a puzzle—a puzzle from the sixties. Gaia is hated by those on the science-technology side, who might have been expected to show some sympathy. It is loved by others, those on the counterculture side, who might have been expected to have been wary, at least of its origins. It was conceived by a scientist who was in the mainstream of science and technology and was supported by a young woman who was starting to gain fame and respect in the scientific community. Why did they think up and stick with such a hypothesis, especially when it was so professionally dangerous? I believe that understanding the present depends upon uncovering the past. Hence, the next chapters take us back in time; when we come again to the present, we will be ready and able to understand the conundrum posed in these last few paragraphs.

3

THE PAGAN PLANET

If you give it a moment's thought, it is blindingly obvious. Suppose you are living in the pre-industrial age, in a country somewhere around the Mediterranean Sea. Life is basically agricultural. Even if you live in a town or a city, we are usually talking about only a few thousand people. You will be aware of the seasons in a way hidden from most of us today. The time of rebirth and growing in the spring. Then the summer and the harvest, leading into the autumn, and finally everything dies back in the winter. Snow and cold, if you live far enough north or in high country. You will realize how important are the springs of fresh water that gush mysteriously from the ground and then pour into the fens and marshes; the streams and rivers—water that can irrigate the fields, that harbors abundant supplies of fish and supports other wildlife, and that provides transport. Rain at the right time and in the right quantities will be of life-saving importance. Drought and flood will be among your biggest fears. You will almost certainly fear lightning and thunder, and if you are close to the sea and sailing on it for a living, you will be wary of the storms that can arise and threaten. You will see the heavenly bodies in a way that most of us cannot today, with our skies made dim by our artificial light. All of this and more. The question is not whether you would think that our planet is a living organism—Mother Earth—but why you would ever doubt this.

Certainly there was no such doubt in Greece five centuries before Christ, where we find the seeds that were to flower into the great philosophical and scientific systems of the ancient world. Thales of Miletus (624–546 BCE), generally thought to be the first of the significant philosophers, argued that ultimately the world is made from one stuff, which he identified with water. According to Aristotle (384–322 BCE), writing some two centuries later, this went along with a belief that everything is infused with life, with soul in some sense. "Certain thinkers say that soul is intermingled in the whole universe, and it is perhaps for this reason that Thales came to the opinion that all things are full of gods" (*De Anima* 411a8–10, in Barnes 1984, 1:655). Earth is part of the universe, and hence it is ensouled, a being that is godlike. This was a conviction shared by Thales's successors, including Anaximander (Thales's student) and Anaximines and Pythagoras (Anaximander's students). It is therefore no surprise that it was also central to the thinking of the key figure in our story, arguably the greatest philosopher of all time, Plato (427–347 BCE). It was he above all who made living worlds part of Western tradition.

PLATO'S PHILOSOPHY

The key source is a later work, the *Timaeus*. To appreciate it fully, we need to grasp general themes in the philosopher's system, starting with the fact that he was in major respects responsible for the belief that the world is no blind, meaningless concoction, simply churning through time without rhyme or reason. He saw in the intricacies and functioning of the physical world (including living organisms) signs of purpose. Such sophisticated complexity, working away, could not be empty chance. There had to be meaning or purpose behind things. Think about why a man grows. "I had formerly thought that it was clear to everyone that he grew through eating and drinking," and that is an end to things (*Phaedo*, 96d, in Cooper 1997, 83–84). But, Plato objected, such an explanation is not sufficient. It is not wrong, but it is incomplete. One must address the question of why someone would grow. Here one must bring in a thinking mind, for without this, one has no way of relating the growth to the end result, the reason for the growth. Our world is structured; it is ordered by an intelligence for good ends. In other words, using a term coined in the eighteenth century, and employing a mode of thinking that we saw in our twentieth-century (Gaia hypothesis) scientists, we encounter in Plato a *teleological* view of things, where objects (including organisms) are considered with respect to their consequences—the eye serves the purpose or end of seeing, the rain serves the purpose of fertilizing the crops—and excellence consists in achieving those consequences. Value arises from ends governing what happens and what one should be striving to make happen. Note that Plato does not suggest that the future influences the present in a regular, causal manner. For him, it is a matter of thinking about the future and letting these thoughts influence what we do now. Given that the teleology is a function of a designing intelligence, something separate from our world itself, it is often said that this is a form of "external" teleology.

This begins to make us think in terms of a god, an ultimate creative intelligence, and as Plato articulated his metaphysical world view—particularly in the middle work, the *Republic*—he gave such an intelligence the fundamental role in his ontology (theory of being). Following the Pythagoreans, Plato was much impressed by mathematics, arguing that there must a world of absolute reality, unchanging, eternal, containing things like numbers and geometrical forms, the objects of mathematics itself. But this world is also stocked with ideal templates, universals—horse, house, human, tree—the things to which we refer when we say not just "Dobbin has four legs" but "A horse has four legs." We humans do not live in this ideal world. We live in the physical world. Our world is not unreal, but it is not fully real. It exists because its objects "participate" in the ideals, and Plato suggested that in some sense this means that the world's objects imitate or are modeled on those ideal templates—known as Forms or (deriving from the Greek word) Ideas. Note that they are not "ideas" in the sense of mental notions; rather, they are what is truly real and objective.

Because our world is not fully real, it can and does change. It is also, as the teleology shows, a world of value. At this point, the theory of Forms was fundamentally explanatory. Plato argued that the Forms are ordered hierarchically. Ultimately they all stem from the One, the Form of the Good—the entity that later generations, especially later Christian generations, identified with the godhead. It is this that gives being and meaning to everything else; it is this from which all other Forms derive, and, at some important level, this ultimate Form is the source of goodness and beauty and all else we find desirable. Scholars suspect that Plato was reporting on a mystical vision, because such visions were known to emphasize the oneness or unity of all existence and to insist on the impossibility of literal description and understanding. Fleshing out this somewhat mysterious doctrine, Plato drew an analogy between the Form of the Good and the sun in our world. Just as the latter gives being and sustenance for us, so the Form of the Good gives being and sustenance in the transcendent world.

With this as background, we turn to the *Timaeus*, a work discussing the origins of the universe and the form that it takes (Johansen 2004; Sedley 2008). Of central importance is the creator, or the craftsman, generally known as the Demiurge. This being is intended to be all wise, in the sense of embodying reason—one might say it is pure Reason. In some respects it is much like the Christian God, although the Demiurge is less a creator in the sense of "out of nothing" and more a designer who works on preexisting materials. But how does it do this? Here Plato's already established philosophy kicks in. The Demiurge models the objects of our world on the Forms. Plato says, "Well, if this world of ours is beautiful, and its craftsman good, then clearly he looked at the eternal model." And with regard to the world itself, it is the Form of the Good that is involved. "Now surely it's clear to all that it was the eternal model he looked at, for, of all the things that have come to be, our universe is the most beautiful, and of causes the craftsman is the most excellent. This, then, is how it has come to be: it is a work of craft, modeled after that which is changeless and is grasped by a rational account, that is, by wisdom" (Plato, *Timaeus* 29a).

Plato does not regard this creation—the universe—as some
dead, lifeless entity. It is a living being with a soul. It has value in its
own right. It is not the Demiurge, but it in some sense embodies
perfection to the extent that we can have this in the physical world.
"Now why did he who framed this whole universe of becoming
frame it? Let us state the reason why: He was good, and one who is
good can never become jealous of anything. And so, being free of
jealousy, he wanted everything to become as much like himself as
possible" (*Timaeus* 29d–e). Because the Demiurge, the designer,
wanted everything to be as good as possible, he worked with the
whole of the physical universe, making order from disorder on the
principle that the former was better than the latter. Clearly the Demiurge, being himself good, had to model things on the best, the
Form of the Good. And this brings in intelligence. The Demiurge
realized that the intelligent is better than the unintelligent and
that the physical on its own cannot supply this. And so straight off
we get a world soul. "Guided by this reasoning, he put intelligence
in soul, and soul in body, and so he constructed the universe. He
wanted to produce a piece of work that would be as excellent and
supreme as its nature would allow. This, then, in keeping with our
likely account, is how we must say divine providence brought our
world into being as a truly living thing, endowed with soul and
intelligence" (30b–c).

We see therefore that, in the world as we see it, Plato integrated
value (it is good) with ends (soul and intelligence). That is what
organisms are all about. "When the maker made our world, what
living thing did he make it resemble?" (*Timaeus* 30c) He had to
go right to the top. In some sense, the world is modeled on the
Form of the Living. "For that living thing comprehends within
itself all intelligible living things, just as our world is made up of
us and all the other visible creatures" (30c–d). It turns out that
the Form of the Living, being the best one can have, is the same
as the Form of the Good. And note that this leads to integration,
to unity: "Since the god wanted nothing more than to make the
world like the best of the intelligible things, complete in every way,

he made it a single visible living thing, which contains within itself all living things whose nature it is to share its kind" (30d–31a). Although in the *Republic* the sun is likened in this world to the Form of the Good in the real world, ultimately it is the universe itself that is modeled on the Form of the Good.

It follows from all of this that certain obligations are laid upon us humans. We may be living on Earth, but in an important sense it is a divine being, over and above being part of the universe and its soul. "The Earth he [the Demiurge] devised to be our nurturer, and, because it winds around the axis that stretches through the universe, also to be the maker and guardian of day and night. Of the gods that have come to be within the universe [other gods include the planets and stars], Earth ranks as the foremost, the one with greatest seniority" (*Timaeus* 40b–c). We must therefore respect Earth and not exploit it. Plato's psychology always stressed balance and harmony. Thanks to the world soul, this holds through the whole universe and not (as for Gaia) just our own Earth; although, as individual Forms emanate from the Good, so the (Gaia-like) individual soul of Earth emanates from the world soul.

AFTER PLATO

Plato's great student Aristotle was as firmly committed to teleology and purpose as his teacher. He spoke of understanding not only in terms of "proximate causes"—the sculptor striking the chisel to chip away at the marble in the act of making a statue—but also of "final causes"—the end or purpose for which the sculptor is making the statue. Yet, for all the overlap, there are significant differences in the thinking of the two philosophers, because Aristotle did not want to make the immediate move from purpose to a godlike designer. Somehow his final cause is to be purely scientific and operating directly in the world. As opposed to Plato's external teleology, his is to be an internal teleology. He raises directly the lack of conscious attention or direction. "This is most obvious in the animals other than man: they make things neither by art nor after enquiry or deliberation." They do things for the sake of ends, but they do these things purely by nature. For instance, "It is both by nature and form that the swallow makes its nest and the spider its web, and plants grow leaves for the sake of the fruit and send their roots down (not up) for the sake of nourishment." Thus, "it is plain that this kind of cause is operative in the things which come to be and are by nature" (Aristotle, *Physics* 199a26–30, in Barnes 1984, 340). There appear to be special forces at work. Note again that these forces are not acting out of the future affecting the present. They exist now but in some sense make reference to or anticipate the future.

How does this all work, and, in particular, where does soul come into the story? As it does for Plato, soul plays a major part in Aristotle's thinking, but, unlike Plato, he does not want to separate it from the body, in the sense that one could imagine the intellectual part of the soul at least existing apart from the body. For Aristotle, body and soul are always in some sense one, with the soul animating and making possible the body. (This position is known technically as *hylomorphism*.) Crucially, however, Aristotle rejected the idea of world souls. Aristotle believed in gods, in particular in a Prime Mover. But, the Prime Mover is not the Christian God—creator of heaven and Earth, who made humans in his own image that he might love us and we might worship him. It is highly unlikely that it is even aware of our existence! For Aristotle, the ultimate god is doing what is appropriate to his nature—a nature obviously modeled on the habits and desires of an upper-class Athenian gentleman—namely, contemplating the intellectual things of life: philosophy, mathematics, and, ultimately, the Forms. From this it is a simple step to the conclusion that the very highest form of being spends its time contemplating its own perfection, because that is the highest thing that can be the object of thought! What it does not do is busy itself with the needs and affairs of the physical world, the universe. Rather, and here teleology kicks in, the Prime Mover acts as a sort of final cause. In some way, because it is ultimate perfection, we and the rest of the universe strive toward it. "There is then something [the heavens] which always moved with an unceasing motion, which is motion in a circle; and this is plainly not in theory only but in fact. Therefore the first heavens must be eternal. There is therefore also something which moves them. And since that which is moved and moves is immediate, there is a mover which moves without being moved, being eternal, substance, and actuality" (Aristotle, *Metaphysics* 1072a21–26). He goes on to say, "Thus it produces motion by being loved, and it moves the other moving things" (1072b3–4). A unified (and unifying) conscious world soul, an *anima mundi* (to use the Latin term), has no place in this picture. Although, obviously, given his thinking about final cause—and given the hylomorphism that mixes up body and soul—Aristotle does not envision a dead, inert world of unthinking material. There were those in antiquity whose thinking pushed this way—the atomists particularly—but most people felt that such thinking was implausible to the point of absurdity. Blind law working on dead matter does not lead to the functioning universe we see so clearly around us. The animation of matter might generally be more vegetative than conscious, but it is there nevertheless.

As we move down through the centuries, world soul thinking—known technically as *hylozoism*—came and went, rarely vanished entirely, and was often in ascendancy, thriving vigorously, often depending on the various enthusiasms for Plato over Aristotle or vice versa. The Stoics, for instance, were very keen on the idea. According to the Roman philosopher and statesman Cicero (106–43 BCE), "A hot and fiery principle is interfused with the whole of nature" (*De Natura Deorum* 2.28, in Hunt 1976, 34–35). In the long run, however, the Hellenistic (Egyptian-born) philosopher Plotinus (204–270 CE), the great interpreter of Plato, was far more influential. His philosophy revolves around three aspects—*hypostases*—of the Form of the Good: the One, the Intellect, and the Soul (Armstrong 1940). The One is the most important, and it stresses the integrated nature of all of reality—a recurrent theme. Nothing exists independently; all is connected. From the One, all other things "emanate" (in the sense of being dependent on it for their existence). "We are in search of unity; we are to come to know the principle of all, the Good and First; therefore we may not stand away from the realm of Firsts and lie prostrate among

her"—swept all before him, especially as interpreted in the
theses produced by the great theologian Thomas Aquinas
274). Again, everything had to be Christianized—we have
eady the differences between Aristotle's Prime Mover and
stian deity—but once this was done, his thought infused
nd the foundations of metaphysics throughout the later
of the Middle Ages. Specifically, his geocentric world pic-
infused with the organic model of understanding. Things
st happen. They make for or serve ends. In Aristotle's
, we seek not just proximate causes but also final causes.
ask where things are going and why, and not just how
ted out. There may not be a case here for a conscious
l, but still there is a sense of life, of soul in some sense,
the whole of reality.

IENTIFIC REVOLUTION

st this background that we must introduce and judge
scientific advances of the sixteenth and seventeenth cen-
talk now of the so-called Scientific Revolution, cover-
eat outburst of empirical inquiry and theorizing of the
and seventeenth centuries, taking us from Nicolaus Co-
announcement in 1543 of his heliocentric theory in *De
ibus Orbium Coelestium* (*On the Revolutions of the Celes-
s*), to Isaac Newton's causal explanations given in 1687
osophiæ Naturalis Principia Mathematica (*Mathematical
of Natural Philosophy*). In physics, we go from the great
ronomer to Tycho Brahe, who measured the heavens
dible precision. His student Johannes Kepler discovered
basic laws governing planetary motion and broke deci-
the ancients' obsession with circles, announcing that
travel in ellipses, slightly elongated figures. Galileo Ga-
d (more probably improved) the telescope and showed
he heavens never before anticipated—mountains on the
moons around Jupiter. He also made great advances
ics, and even today we use his basic laws to understand

the lasts: we must strike for those Firsts
which are the lasts." Hence, "Cleared
toward the Good, we must ascend to
selves; from many, we must become o
knowledge of that which is Principle
neads, 6.9.3).

Like Plato's, Plotinus's whole univ
nected life. What is genuinely historic
Plotinus's interpretation of Plato tha
more) the Western world knew, not
work. Indeed, right into the Middle A
Plato was known, and, significantly, tl
it was to be used, and people did war
had to be Christianized. For instance
about the Demiurge just designing;
ing. Also, some interpretation had to
support of the *Timaeus* for the cruc
Trinity. Fortunately, the medieval *T*
a commentary by a fourth-century
fered such an interpretation. The D
the Father. The set of Forms from
models the world is God the Son.
that pervades all physical reality is Go
brilliant resolution of the issue, expla
do and how God is ever present (I
our planet having value in its own r
life of our planet is in some sense io
of the Almighty Creator. The origin
but it was a paganism that many at
Christianity—when the Church ha
own understanding of the essentia
to their being imposed by the out
felt lent itself to transformation int
teenth century came the rediscov
translation from Islamic texts and
the libraries of Europe. He—no na

philoso
great sy
(1225–
noted a
the Chr
science
centurie
ture wa:
do not
languag
We mus
they sta
world sc
pervadin

THE S

It is aga
the great
turies. W
ing the g
sixteenth
pernicus'
Revolutio
tial Spher
in his *Phi*
Principle
Polish as
with incr
the three
sively wit
the plane
lilei devis
details in
moon an
in mecha

moving bodies. Finally came the great synthesizers, first the French genius in philosophy, mathematics, and science, René Descartes. He postulated his vortex theory, in which suns are at the centers of celestial whirlpools—all matter ultimately is tiny particles, atoms— and planets are carried along in their flow. Then Isaac Newton postulated his three laws of motion and his law of gravitational attraction—all bodies attract each other with a force inversely proportional to the square of the distance between them. He showed that motions in the heavens, discovered by Kepler, and motions here on Earth, discovered by Galileo, are governed by one and the same set of principles. No longer is the universe—as it was for Plato and Aristotle—divided into a perfect area above and an imperfect area below. Scientifically they are all one and the same.

Many others contributed to this revolution in different fields— Harvey, with his discovery that the heart is a pump, and Gilbert, with his discoveries about magnetism, to name but two. There were also theorcticians or philosophers of the Scientific Revolution, notably the English statesman Francis Bacon and, later in the seventeenth century, a man justly remembered for his great discoveries in chemistry, Robert Boyle. They saw (rightly) that the essence of the revolution was a change in perspectives, a change in metaphors. At the beginning of the sixteenth century, people agreed with the great Greek philosophers that the world was organic in some sense. It is appropriate, therefore, to ask about final causes. Aristotle's physics was end-directed, teleological—one tried to understand motion in terms of the natural resting places for the elements (earth at the center and then, in order, water, air, and fire). The celestial physics of the ancients was no less end-directed, with everything being referred to final causes. At the end of the seventeenth century, the mechanical metaphor ruled. The world was seen as a machine, as in some sense clocklike. Matter was no longer living, but dead, inert, acted upon by external forces that cause it to go on moving and persisting without beginning and without end—above all without purpose. Values have to be imputed, not discovered.

Descartes is the most significant thinker here, for he articulated

a philosophy that expressed this shift (Garber 1992). His fundamental assertion was to separate mind and matter into two kinds
of ultimate substance, *res cogitans*, or thinking substance(s), and
res extensa, or material or extended substance(s). Thought—ideas,
inferences, emotions, and the like—is thinking substance. Unlike
Plato's mind-stuff, it has no dimension. Physical objects—planets,
pendulums, plants, and, notoriously, even animals—are extended
substance, and necessarily they cannot be alive in the sense of
things with the other kind of substance, thinking substance. But
what they can do is go through the motions that earlier generations had thought indicative of some kind of special life force.
In the *Discourse on Method*, Descartes discussed Harvey's work,
showing that the heart is a pump (a machine) and explaining the
similar mechanistic functioning of other bodily parts. "This will
hardly seem strange to those who know how many motions can
be produced in automata or machines which can be made by human industry, although these automata employ very few wheels
and other parts in comparison with the large number of bones,
muscles, nerves, arteries, veins, and all the other component parts
of each animal" (Descartes 1964, 41).

It was Boyle, much impressed by those wonderful timepieces in
medieval churches that not only tell time but show the motions
of the planets and have moving figures that perform on the hour,
who drummed home the clock image. He would not have us think
of the world as crammed with life forces, forever working to keep
things going. Referring specifically to a device built in the late sixteenth century, he argued that the world is "like a rare clock, such
as may be that at Strasbourg [fig. 2], where all things are so skill-
fully contrived that the engine being once set a-moving, all things
proceed according to the artificer's first design, and the motions of
the little statues that at such hours perform these or those motions
do not require (like those of puppets) the peculiar interposing of
the artificer or any intelligent agent employed by him, but perform
their functions on particular occasions by virtue of the general and
primitive contrivance of the whole engine" (Boyle 1996, 12–13).
This was not a move to atheism. Those who believed that the

Figure 2. The automatic clock in Strasbourg Cathedral, built in the 1570s. (Woodcut by the sixteenth-century Swiss illustrator, Tobias Stimmer.)

world was the creation of a caring God immediately inferred that the world was in some sense an artifact designed and made by God. If the world seems machinelike, it is because it is machinelike! It is God's machine. Descartes made this very clear. Following the passage from the *Discourse on Method* quoted above, he wrote that we should think of the human body "as a machine created by the hand of God, and in consequence incomparably better designed and with more admirable movements than any machine that can be invented by man" (Descartes 1964, 41).

But, ideas have their own momentum. Once the machine metaphor had been promulgated, scientists increasingly found that it simply was not profitable to ask questions (as scientists) about why God had made the machine. "If an omnipresent God was all spirit, it was the more easy to think of the physical universe as all matter; the intelligences, spirits and Forms of Aristotle were first debased, and then abandoned as unnecessary in a universe which contained nothing but God, human souls and matter" (Hall 1954, xvi–xvii). Scientists focused on the workings of constant laws, and in the physical world questions about final cause came under increasing attack. They were neither useful nor necessary. Francis Bacon humorously likened final causes to vestal virgins, decorative but sterile. Descartes was no less contemptuous. How can we ever be truly certain as to God's intentions? We should not be so arrogant as to presume we can ferret out His ways and His ends (Descartes 1644, 1:28). The consequence of such thinking was that, by the end of the seventeenth century, as far as science was concerned, in the words of another of the great historians of the Scientific Revolution, God was regarded as "a retired engineer" (Dijksterhuis 1961, 491). In such a world view, there was little room for souls, either in the universe as a whole or in planets like Earth in particular. Apparently, hylozoism was well and truly dead.

PLATONISM VIVIDUS

Yet the history of the Scientific Revolution is more complex and more interesting than you might expect. It is just not true that by the end of the seventeenth century the machine metaphor had conquered all and that everyone saw the physical world as nothing but blind matter in endless motion. Platonism persisted, and with it the idea of a world soul. All scholars of the period now recognize the great influence of Platonic thinking on Copernicus. His move to make the sun central is very much the move of a Platonist (and, more primitively even, a Pythagorean) who sees the sun as having a crucial role in the ontology of this world, corresponding to the Form of the Good in the world of Ideas. Even more striking is Johannes Kepler. At times he sounds like the archetypal new scientist, putting world souls and the like firmly behind him. The clock metaphor rules triumphant: "It is my goal to show that the celestial machine is not some kind of divine being but rather like a clock." But he convinced no one, including himself. Overall, he was fanatical in his Platonism. "By the highest right we return to the sun, who alone appears, by virtue of his dignity and power, suited for this motive duty and worthy to become the home of God himself, not to say the first mover" (Burtt 1932, 48, quoting an early fragment).

Kepler was an empiricist, fitting theory to the facts, and this is the foundation of his claim to fame. However, this is but part of a more complex program, one setting out explicitly to find the hidden mathematical properties of a sun-centered universe. Just consider his famous (or infamous) construction for measuring the distances of the planets from the sun. It is not a simple matter of measurement and leaving things at that. In the *Timaeus*, Plato made much of the "perfect forms," that is, the (five kinds of) three-dimensional objects with identical faces, arguing that they are the ultimate constituents of physical reality. Kepler seized on this, arguing that these forms have even greater importance, for they are the basis of the way that the Great Geometer in the Sky constructed the universe: it is not just chance that there are six and only six planets, including Earth (fig. 3). But even this was not enough for Kepler. "The view that there is some soul of the whole universe, directing the motions of the stars, the generation of the elements, the conservation of living creatures and plants,

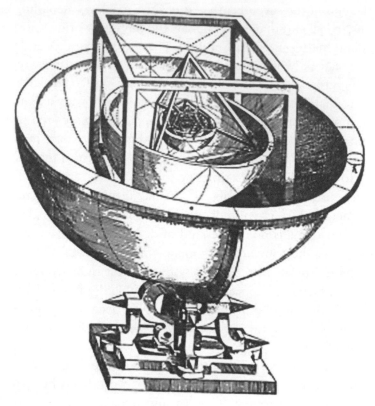

Figure 3. The perfect solids shown as fitting within each other, thus yielding the distances of the planets from the sun and showing how deeply Kepler's thinking was embedded in the Pythagorean/ Platonic tradition. (From Kepler's *Mysterium Cosmographicum*, 1596.)

and finally the mutual sympathy of things above and below, is defended from the Pythagorean beliefs by Timaeus of Locri in Plato" (*Harmonice Mundi*, 1619, in Kepler 1977, 358–59). Having given a Christian blessing to such speculation, with an enthusiasm that would not have been shared either by the Greek philosopher or the preacher from Galilee, Kepler explored in some detail the analogies between the functioning of Earth's own soul and more familiar bodily workings, arguing that "as the body displays tears, mucus, and earwax, and also in places lymph from pustules on the face, so the Earth displays amber and bitumen; as the bladder pours out urine, so the mountains pour out rivers; as the body produces excrement of sulphurous odor and farts which can even be set on fire, so the Earth produces sulphur, subterranean fires, thunder, and lightning; and as blood is generated in the veins of an animate being, and with it sweat, which is thrust outside the body, so in the veins of the Earth are generated metals and fossils, and rainy vapor" (363–64).

Let us move the clock forward rapidly. Mechanism may have been the philosophy of the day, but the rise of humanism, with its interest in the ancients, and a corresponding willingness to learn the appropriate languages, especially Greek, helped Platonism to thrive. Kepler was not alone in harking back to earlier themes; the University of Cambridge, especially, housed a school of theologians and philosophers who were deeply influenced by the great philosopher. As in earlier times, scholars identified key Christian concepts with key Platonic concepts, especially with regard to Plotinus's trilogy of the One (the Form of the Good), the Intellect (the full array of Forms, somehow bound up with the Demiurge), and the Soul (the life force, associated with desire). These were all brought together and (as was traditional) associated with the Trinity. And with this, as in earlier times, we get living matter and world souls. The philosopher and theologian Henry More, the most important and interesting of the Cambridge Platonists, felt that some kind of vital force—a Spirit of Nature—must exist to keep things going, something acting all of the time (Hall 1996). This vital force must pervade the universe, and thus More broke with an earlier enthusiasm for Descartes (and sided with Plato) in thinking that mind or spirit had dimensions. He thought it existed in space, just as does matter. It is not so much that this Earth of ours is an organism as such, but that the whole of the universe is living—not necessarily in a conscious way, but in a way that animates and moves brute matter.

This sort of thinking did not go unchallenged. Robert Boyle particularly would have none of it. You simply did not need things like the Spirit of Nature to do science. They were at worst wrong, and at best redundant. "The phaenomena I strive to explicate may be solved mechanically, that is, by the mechanical affections of matter" (Boyle 1996). All one had to do was explain things in terms of motion, size, mass, shape, and the like. Doing this was enough. Or was it? What about gravity itself, the central notion of the physical theorizing of Isaac Newton? In supposing that bodies can affect each other across spaces—action at a distance—there is the scent (the Cartesians said a stench) of something like spirit. This was the epitome of what good French mechanists thought was barred by the new philosophy. Dare one suggest that Newton had fallen under the Platonic spell of his good friend Henry More? But More and his fellow Platonists were not alone in such thinking. There were scholars on the Continent who would not be forced into Cartesian straitjackets—for example, the Dutch philosopher Spinoza, who identified God with Nature, and the German philosopher Leibniz, who thought that all of reality was made up of self-contained "monads."

Above all, hovering over the whole discussion was the organic world itself. It was all very well to sneer at final causes, but in the world of plants and animals, they remained pressing. As Robert Boyle himself pointed out, final-cause thinking was still needed with regard to the living world. No one could deny that the eye was made for the end of seeing. That was its purpose. One of the organisms that Boyle discussed in some detail was the bat: "Though bats be looked upon as a contemptible sort of creatures, yet I think they may afford us no contemptible argument to our present purpose" (Boyle 1966, 194). He discussed the membranes between the digits that made the wings, the hooks on the wings for holding onto trees, the teeth for chewing, the internal organs of females for bringing forth live offspring, and even the "dugs, to give suck to her young ones" (196). She is even restricted to two teats, because that is the number of offspring that she has.

Of course, theologically, one can talk about these final causes. One can ask questions about the intentions of God and so forth— how things have value because God intended things this way. But such inquiry is not admissible in scientific discourse: "This is not the proper task of a naturalist, whose work, as he is such, is not so much to discover *why*, as *how*, particular effects are produc'd" (Boyle 1966, 229–30). But how then does the scientist account for the need to consider final causes in the biological world when they have been expelled from the physical world? Right through the seventeenth century, scientists used final causes in their study of organisms and worried about their underpinning. How can one get a scientific account without putting it all down to God? Grappling with this problem completes our move from the happy, sunlit groves of British philosophy to the dark and thorny thickets of German idealism.

FRIEDRICH SCHELLING AND *NATURPHILOSOPHIE*

The key linking figure in our story is the great modern philosopher Immanuel Kant. In major respects Kant accepted a mechanistic worldview, arguing that ultimately this is a function of our interpretation of some kind of independent reality, the "thing-in-itself" (*Ding an sich*). In Kant's language, ultimate being is the noumenal world, as opposed to the phenomenal world that we encounter directly and that we ourselves structure in some respects. Kant soon realized, however, that his system might work for the inorganic world but that the world of organisms posed new challenges. Organisms show an organization, an integration, that cannot be entirely explained through mechanical accounts. Then how do we explain this organization, this working to final cause? It is obvious that Kant, like Boyle, wanted to take it all back to God, but, again like Boyle, he was adamant that such an explanation could have no place in science. Hence, somewhat uneasily, he argued that we can look at the biological world *as if* it were designed, without committing ourselves to the reality of such design. One consequence of this is Kant's belittling of the status of biology as a possible full science. "We can boldly say that it would be absurd for humans even to make such an attempt or to hope that there may yet arise a Newton who could make comprehensible even the generation of a blade of grass according to natural laws that no intention has ordered; rather, we must absolutely deny this insight to human beings" (Kant 2001, 270). The world may be a machine, but certain aspects of it will forever elude our understanding.

Kant's immediate successors, especially in his home country, seized at once on the weaknesses in his system. Above all, the thing-in-itself came under severe scrutiny and criticism. It had to go, and thus German philosophy almost naturally took a turn toward idealism. The only real was the ideal! This particularly was the stance of Johann Gottlieb Fichte. But obviously this was hardly satisfactory. To put it in modern terms, does one really want to say that the only reality of dinosaurs is as figments of our imagination? Do they really have no existence outside our minds?

Fichte's sometime follower and then opponent, Friedrich Wilhelm Joseph Schelling (1775–1854), saw the problem. His solution made immediate reference to Spinoza. God or Nature: they are one. In some way, it is not a matter of the real—the table or the dinosaur—becoming part of the ideal, the imagination, but of the real and the ideal being one and the same. "In that I envisage the object, object and idea are one and the same. And only in this inability to distinguish the object and the idea during the envisioning itself lies the conviction, for the ordinary understanding, of the reality of external things, which become known to it, after all, only through ideas" (Schelling 1988, 12). But there was a deeper and earlier influence. The nineteen-year-old Schelling had penned a fifty-eight-page essay on Plato's *Timaeus*, and the overall effect was massive and long-lasting (Beierwaltes 2003). In the Platonic Ideas, Schelling saw not only ultimate objective reality but also ultimate subjectivity, for they have their being in the mind of the Demiurge. "The key to the explanation of the entirety of the Platonic philosophy is noticing that Plato *everywhere carries the subjective over to the objective*" (Schelling 2008, 212; his italics). For Schelling, then, everything ultimately resides in what he and other idealists call the Absolute—their equivalent of the Platonic creator and designer. At this point we begin to see another significant move beyond Kant. Schelling rejected the judgment that accepting the end-directedness of organisms meant that biological understanding was inferior. It could not be that the only good explanations were mechanistic explanations. If one side of the Absolute is human awareness and understanding, the subjective, then at some level this must be reflected in the other side, the objective. In other words, the world in any sense must be essentially organic, with final cause an essential part of it. "Even in mere organized matter there is *life*, but a life of a more restricted kind. This idea is so old, and has hitherto persisted so constantly in the most varied forms, right up to the present day—(already in the most ancient times it was believed that the whole world was pervaded by an animating principle, called the world-soul, and the later period of Leibniz gave every plant its soul)—that one may very well surmise from the beginning that there must be some reason latent in the human mind itself for this natural belief" (Schelling 1988, 35). There is indeed a reason for this belief. "The sheer wonder which surrounds the problem of the origin of organic bodies, therefore, is due to the fact that in these things necessity and contingency are most intimately united. *Necessity*, because their very *existence* is *purposive*, not only their form (as in the work of art), *contingency*, because this purposiveness is nevertheless actual only for an intuiting and reflective being" (35).

A term was now introduced that is still popular in some circles today. Apparently, because of the notion of purpose, "the human mind was very early led to the idea of a *self* organizing matter, and because organization is conceivable only in relation to a mind, to an original union of mind and matter in these things" (Schelling 1988, 35). Self-organization! The world is something that produces itself, has its developing powers inside, as an unfurling organism is driven by forces within rather than without. One goes from the simple to the complex, from the undifferentiated to the highly differentiated. "Nature should be Mind made visible, Mind the invisible Nature. Here then, in the absolute identity of Mind *in us* and Nature *outside us*, the problem of the possibility of a Nature external to us must be resolved. The final goal of our further research is, therefore, this idea of Nature; if we succeed in attaining this, we can also be certain to have dealt satisfactorily with that Problem" (42). It all comes together. Schelling's philosophy of nature, *Naturphilosophie*, did not come out of nowhere. It is almost self-evident that in major respects, going back through Fichte, Kant, and Spinoza, his philosophy stood in a line that began with the thinking of Plato—and not just any part of Plato, but the Plato of the *Timaeus*. We see this not just in Schelling's collapsing of the distinction between objective and subjective and making central the organic metaphor (or literal claim, if you prefer), but also in his insistence that everything is ultimately one and that this one (or Absolute) is essentially ideal and the source of all being. This is deeply Platonic.

Now, having entered the nineteenth century, with the coming of a philosophy like Schelling's, representative of what we now call the Romantic movement, we see that his kind of thinking was catching fire, inspiring thinkers and writers in both the arts and in science (Richards 2003). Matter was seen as no dead pile of stuff, Descartes' *res extensa*, but as something pulsating with life through and through. This is just the time when people like Ben Franklin were pushing forward the science of electricity, which suggested that forces or invisible fluids were at issue here, and others were showing that life itself involves not just fluids but electrical discharges, such as those needed for the functioning of muscles. Even physics was being co-opted into this movement, for the Jesuit Roger Boscovitch argued—in a move endorsed by Kant in his *Metaphysical Foundations of Science*—that matter itself can be reduced to opposing forces, those pushing out and those pulling in. It pushes out as you try to penetrate it, but at the same time it pulls back in, or it would simply diffuse throughout the universe: "Repelling force belongs to the essence of matter as much as attractive force does—the two can't be separated in the concept of matter" (Kant 2009, 30).

These ideas were not confined to science and philosophy. Writers and painters saw life throughout all creation. And life involved growth, becoming rather than just being. Nothing stands still. The world throbs with vitality, meaning, and purpose—and all is essentially and eventually (in a Platonic way) one. The poet Goethe epitomized this movement, writing about it in verse and prose and, not hesitating to consider the sciences, holding forth on such matters as the nature of light and color; the supposed ways in which the vertebrate skull is made from modified identical parts, originally vertebrae; and the development of plants from a basic, shared archetype—the *Urpflanz* (fig. 4). And Goethe was not alone, nor was the Romantic movement confined to Germany:

Figure 4. Goethe's *Urpflanz*, the archetype of all plants. From top to bottom, the flower parts are: pistil (female reproductive part), stamens (pollen-producing reproductive organs), corolla (petals), and calyx (sepals). (From J. W. von Goethe, *The Metamorphosis of Plants* [*Versuch die Metamorphose der Pflanzen zu erklären*] [Gotha: Carl Wilhelm Ettinger, 1790].)

And I have felt
A presence that disturbs
    me with the joy
Of elevated thoughts; a
    sense sublime
Of something far more
    deeply interfused,
Whose dwelling is the
    light of setting suns,
And the round ocean,
    and the living air,
And the blue sky, and in
    the mind of man,
A motion and a spirit,
    that impels
All thinking things, all
    objects of all thought,
And rolls through all
    things.
(William Wordsworth,
Tintern Abbey, 1789)

The coming of the machine metaphor may have been the

most important event in modern science, but ideas about world souls, integrating forces, and living matter still had their attractions. They persisted right through the Scientific Revolution and well beyond. We examine next how they have fared in the years since.

4

MECHANISM

We start the move now from the beginning of the nineteenth century up to the present, up to the arrival of the Gaia hypothesis. The violent disagreements sketched in chapter 2 suggest strongly that people were coming to the table with very different world pictures, different philosophies of nature. This and the next two chapters identify and discuss these various strands, showing how they fared and developed through the nineteenth and twentieth centuries. Although I do not implicitly suggest that history shows a progress from the past to the present, from the less to the more, from the worse to the better, the discussion is shaped by and focused on the understanding that we can hope to bring to the Gaia controversy. To this end, we begin with the story of mechanism in these years. If it was the dominant world picture of the working scientist, how then did it deal with the issues that are involved in Gaia? Two pertinent areas of inquiry stand out as particularly significant. First there are the earth sciences, specifically geology. Gaia is about Earth, and so the history of work in this area is an obvious choice. Second there is the science of organisms, biology, especially, at this point, evolutionary biology. Gaia's innovative emphasis on the living world as a vital part of the picture requires discussion of this area of science as well.

Since I defer questions about limits and alternatives to the following two chapters, our focus here is on discoveries and theorizing in these areas of science, especially on the extent to which mechanism functioned as the guiding principle of inquiry and on the successes of using this approach. Let me first make clear precisely what it means to say that mechanism functioned as the guiding principle of inquiry (Ruse 2005). Ultimately, mechanism means in some way being governed by the metaphor or model of a machine. We are looking at the material world as if it were a machine. But what does this mean exactly? The basic idea is that the world (meaning the whole of the physical world) is bound by constant, general laws. As the hands of a clock (a machine) keep going round and round because of the laws of mechanics, so the world as a whole keeps moving as it does because of all of the laws of nature. A mechanist in this sense sees a cannonball traveling in a parabola because of Galileo's laws, and a stick seeming to bend when half-immersed in water because of Snell's law, and hydrogen burning in air and leaving a wet residue because of the laws of chemistry. What we have here is the minimalist notion of the metaphor, for God is a "retired engineer"; things do not combine to function for some end, intentional or otherwise. The notion is minimalist but powerful: teleology is expelled. Values are excluded.

However there is a stronger sense that does include the idea that things can work together in a more recognizably machinelike fashion. To quote again the great historian of the Scientific Revolution, physicists often tried "to form as concrete a picture as possible of the physical reality behind the phenomena, the not directly perceptible cause of that which can be perceived by the senses; they were always looking for hidden mechanisms, and in so doing supposed, without being concerned about this assumption, that these would be essentially the same kind as the simple instruments which men had used from time immemorial to relieve their work, so that a skilful mechanical engineer would be able to imitate the real course of the events taking place in the microcosm in a mechanical model on a larger scale" (Dijksterhuis 1961, 497). This sense of mechanism therefore stressed real machines. The aim was to show that the world really does work like the machines that we know and have designed and created. It is not governed solely by abstract laws of motion. Note, however, that thinking in terms of mechanical models, of real functioning machines, does not plunge us right back into Greek final causes. No one (today) is saying that these models were designed in any sense in order to produce their ends. It is rather that they work in a coordinated way.

This second sense of mechanism has always been considered distinctly British (and perhaps later American), though this has not necessarily been a cause for praise or imitation, as shown by this sneering comment by the great French scientist-philosopher Pierre Duhem about *Modern Theories of Electricity* by Sir Oliver Lodge. "Here is a book intended to expound the modern theories of electricity and to expound a new theory. In it there are nothing but strings which move round pulleys which roll around drums, which go through pearl beads, which carry weights; and tubes which pump water while others swell and contract; toothed wheels which are geared to one another and engage hooks. We thought we were entering the tranquil and neatly ordered abode of reason, but we find ourselves in a factory" (Duhem 1954).

The obvious reason for the dominance of such thinking in British science by and after the eighteenth century was the beginning and ongoing success of the Industrial Revolution. As never before, the British were harnessing the forces of nature, and science and technology were closely intermingled. One major consequence for our story is that the kinds of machines at which the British were excelling began to become significant in their metaphors (Mayr 1989). Boyle emphasized the metaphor of the clock, which persisted at least to the beginning of our time period. Archdeacon William Paley makes much of the clock in his celebrated articulation and defense of the argument from design (for the existence of God) in his *Natural Theology*, published in 1802. But there were always critics of the clock metaphor. Shakespeare, in *Love's Labor's Lost*, likens a woman to a timepiece, always breaking down, "And never going aright, being a watch." More positively, as we enter the eighteenth century, great new inventions begin to appear. One of the most brilliant was the Newcomen engine, a device that harnessed the power of steam and was used to pump water out of mines (fig. 5). Also significant were various kinds of feedback mechanisms, particularly the centrifugal governor introduced at

Figure 5. The Newcomen engine. Steam is blown up into the cylinder (*C*). When it is full, a valve closes, and a squirt of cold water is introduced, causing condensation and a vacuum that pulls down the beam. When the beam is all the way down, the valve is opened, and steam is readmitted, forcing the beam to move back up again. The whole cycle is repeated about twelve times a minute. (From Chambers's, *Cyclopaedia*, 1786.)

the end of the eighteenth century onto the successor to the Newcomen engine, the Boulton-Watt engine, named after its two inventors, Matthew Boulton and James Watt (fig. 6).

Since the sixteenth century, mechanism has often been linked with another trend in science, namely, reductionism. Like mechanism, this is a notion with several meanings; but the sense being referred to here is that of trying to explain the larger in terms of

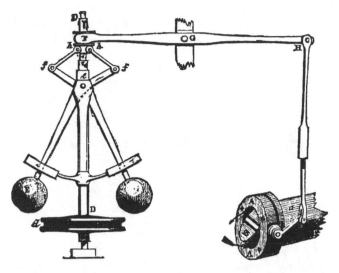

Figure 6. The Boulton-Watt regulator. As the machine speeds up, the central rod spins more rapidly, and centrifugal force makes the balls fly outward. This works to move a valve that reduces the flow of steam, so that the machine slows down. A balance is reached between the speed of the machine and the opening of the valve. (From D. K. Clark and J. Sewell, *An Elementary Treatise on Steam and the Steam-Engine Stationary and Portable [Being an Extension of The Elementary Treatise of Mr John Sewell]* [London: Crosby Lockwood and Company, 1879].)

the smaller (Nagel 1961). It is fairly easy to see how mechanism and reductionism get linked up, especially when we think of the machine metaphor in the second, stronger sense described above. If you want to find out how a machine works—for example, the diesel engine in your truck—your best strategy is take it apart, trying to ferret out the purpose of each part and what it contributes to the larger system. Remember that Descartes, the mechanist's mechanist, had his vortex theory—a theory of the universe according to which planets were pushed around suns by little corpuscles, ever-divisible bits of *res extensa* (fig. 7). Hence, it is often thought that the best way to realize the mechanistic program is to pursue a reductionist strategy.

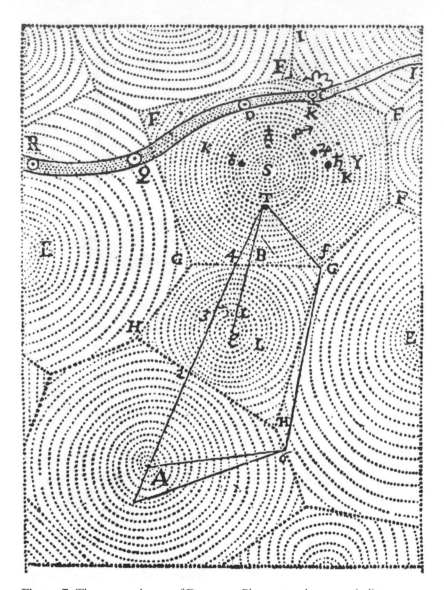

Figure 7. The vortex theory of Descartes. Planets are shown as circling around their suns rather like ships caught in a whirlpool. For Descartes, matter is everywhere, and the dots represent very tiny pieces of matter (corpuscles) that push or drag along larger chunks of matter, particularly the planets. (From René Descartes, *Oeuvres de Descartes*, vol. 11, *Le Monde*, ed. C. Adam and P. Tannery. 12 vols. [Paris: Léopold Cerf, 1897]; written around 1630, first published in its entirety, 1677.)

GEOLOGY

The beginning of the nineteenth century is a good time to take up the history of geology, because people realized then that they had to get serious about making geology more than just a collection of ideas and techniques, where sophisticated knowledge and practices were often carefully guarded and seldom systematized (Oldroyd 1996). Thanks particularly to the Industrial Revolution, there was a high premium on proper understanding of our planet. Factories needed ores and minerals—iron, copper, and tin—and, above all, fuel. Wood was practically exhausted, so there was great demand for coal, which yielded much more heat. All of these demanded mining expertise and geological knowledge. No one wanted to drop a shaft without some assurance of finding coal seams or iron ore deposits. Improvements were also needed in transportation. Macadam surfacing for roadways was invented in the second quarter of the nineteenth century, and railways were beginning to be established. Before these, canals crisscrossed the British countryside, linking the growing urban centers. All of these required geological knowledge, from getting the right stone from quarries to plotting the most efficient cuts, paths, and tracks. To decide whether to tunnel through a hill or go around it, one had to know whether the hill was made of soft sandstone or hard granite, and a geologist was needed to determine its composition.

In many respects, geology is a very empirical science. A huge amount of effort was required to produce maps and to discover the locations of particular rocks and minerals. It was not a question of having no theory until the mapping was done—Charles Darwin rightly said that no observation is of any value without a theory that it is testing—but that the gathering the information was both intellectually and commercially crucial and that it was a massive task. As and when this work was done, what of theory? The theory of major interest to us was developed in the late eighteenth century by the Scottish physician and farmer, James Hutton (Repcheck 2003). Vulcanism argued that Earth's center is hot liquid rock and that the surface pressure from this and its occasional breaking through and bubbling out are the chief causal factors of Earth's geological features. Hutton did not ignore weathering and erosion. He argued that once the underlying liquid had burst through and solidified into rock, it began to be worn down by the forces of nature, and this would lead to sedimentary rocks. He predicted that sometimes we should see sediment on top of rock from underground, and he and his supporters hailed the discovery of such "unconformities" (as they are called in the trade) as major evidence in favor of his theory.

Hutton is best known for his demand that Earth be considered indefinitely old, since it goes through cycle after cycle of eruption and erosion. "We find no vestige of a beginning, no prospect of an end" (Hutton 1788, 304). This points significantly to aspects of mechanism that are of interest to us. Hutton was clearly a mechanist in the first basic sense of using physical laws to explain phenomena. He was also a mechanist in the second sense of thinking in terms of actual machines or models. This is not surprising. Not only was Hutton writing in Britain at the end of the eighteenth century, when the Industrial Revolution was roaring ahead, but he had toured the English Midlands and conversed with and seen the work and results of inventor-industrialists such as James Watt. Watt was right in the thick of designing and making steam engines, which were intended to go on indefinitely, given the needed fuel. Since the major use of these steam engines was in mining, pumping out water, the image of such machines was readily transferred into the world of geological theory (Rudwick 2005). Hutton's primary essay began thus: "When we trace the parts of which this terrestrial system is composed, and when we view the general connection of those several parts, the whole presents a machine of a peculiar construction by which it is adapted to a certain end. We perceive a fabric, erected in wisdom, to obtain a purpose worthy of the power that is apparent in the production of it" (Hutton 1795, sec. 1, 1). Essentially, the world is just like a Newcomen steam engine. The heat from below pushes things up, as in the engine; then it all starts to cool, as in the engine; and finally through the force of gravity things start to fall back down, as in the engine. And like a good steam engine, so long as there is a steady supply of fuel, it just goes on and on, indefinitely.

FROM THE NINETEENTH TO THE TWENTIETH CENTURY

The successor to Vulcanism was the uniformitarianism of Charles Lyell (and soon of his ardent follower, Charles Darwin). Much in the spirit of Hutton, Lyell's great work *The Principles of Geology: Being an Attempt to Explain the Former Changes in the Earth's Surface by Reference to Causes now in Operation* (1830–33) argued that time was the key to geological understanding. As Lyell's subtitle made clear, given enough centuries, all geological change can be explained by referring to everyday processes such as rain and weathering, erosion and deposition, volcanoes and earthquakes. Inventively, Lyell had a kind of waterbed view of the Earth's surface (fig. 8). As one part gets squashed down, through deposition of new strata, another part springs upward to compensate. There is feedback going on here, and a significant consequence is that the relative surfaces of land and sea are always changing. (In this sense Lyell's theory resembles plate tectonics, to be discussed shortly; but Lyell's land always moves up and down, never laterally.) With this view of land and sea, Lyell was able to introduce his "grand new theory of climate." He argued that, because such things as oceanic currents can alter dramatically the climate of the lands they touch, the temperature of any part of Earth has less to do with its distance from the equator than with the relative distributions of land and sea. The Gulf Stream warming Britain is a case in point. Lyell thus avoided the generally held opinion that Earth is cooling. We have a steady state.

All of this is very much part of the mechanistic approach to nature that sees everything as governed by constant law. Lyell was always ready to invoke the machine model to explain various phenomena. That is what drives climate, and, as he filled in the details, the language of machines crops up repeatedly. In a later edition of the *Principles* (9th ed., 1854), in the discussion of earthquakes,

Figure 8. The frontispiece of the first volume of Charles Lyell's *Principles of Geology* (1830), illustrating his theory that Earth is in constant flux, with some parts subsiding and other parts being elevated. The columns are eroded only on their upper parts, suggesting that at some point after the building was completed the land sank, and the building was partially submerged, and then at some later time, it rose again to its previous level. Ultimately everything remains in place.

Lyell wrote that they "constitute an essential part of that mechanism by which the integrity of the habitable surface is preserved, and the very existence and perpetuation of dry land secured" (492–93). Such dual mechanistic thinking persisted, and still persists, in geology. To see that today's geologists fit firmly into the tradition that emerged from the Industrial Revolution, let us now leapfrog across a hundred years. Bearing in mind Lyell's theory that (in some parts of the world) land is always being uplifted as (in other parts of the world) land is subsiding, let us look at a geological theory of mountain building, published just before the Second World War, and now much regarded as an anticipation of the mid-twentieth-century, all-inclusive synthesis theory of plate tectonics.

David Griggs (1939), an American geologist, began to consider "subcrustal convection currents." By now scientists knew that Earth had a central solid core, next fluid, and finally a solid shell (crust) at the surface. Griggs hypothesized that beneath this crust, thanks to the heat down at the core (residual and fueled by radioactive decay), there are the kinds of convection currents that we see in a kettle of water heated on a stove. The heat causes the liquid at the bottom to expand; because it is less dense than the cooler liquid above, the bottom liquid starts to rise, driving the cooler top liquid down. This is in turn heated and rises, displacing the now cooling surface liquid, resulting in a constant circling. Griggs pointed out that mountains are apparently not being built constantly, but arise in occasional spasms, as it were. He set up a model showing how, with a suitably placed valve on the liquid going up, this could occur (fig. 9).

Griggs now applied this kind of thinking to the real world, although again he was working with models, mechanical contrivances that he actually built to demonstrate how things could proceed. First he showed how things would work with two convection currents going in opposite directions. The crust would get sucked down between them and would therefore tend to thicken (fig. 10). Subsequent sedimentation could add to this thickening. If the circulation slows, then by a kind of Lyellian process, we get a regaining of balance: the expanding, heated, liquid rock down below leads to a rise of the land into mountains. There is no need of a

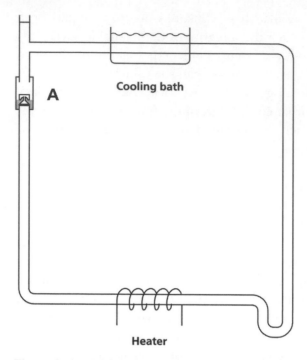

Figure 9. A model to show how heating and cooling can cause the ongoing circulation of liquids. Placing a valve at point *A* allows for periodic variation in the rate of circulation; rapid circulation can be followed by virtual inaction. This simulates the hypothesized circulation of Earth's mountain-building fluids. (Reprinted by permission of the American Journal of Science, from D. Griggs, A theory of mountain building, *American Journal of Science* 237 [1939]: 611–50, fig. 11, 632.)

physical valve to make this happen. If, as seems likely, the circulating fluid is sufficiently viscous, it acts as its own valve—the cooler fluid above makes the hotter liquid below back up until the pressure of the hotter liquid makes things move again. Second, Griggs showed how things work with only one side circulating (fig. 11): the land on the noncirculating side crumples up, and we get mountains that way too. Figure 12 diagrams the whole process.

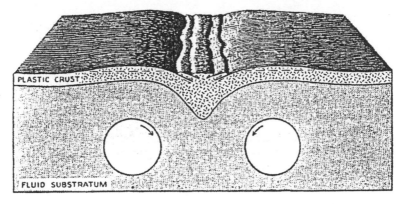

Figure 10. A stereograph of a physical model (involving rotating drums) built by Griggs to simulate the formation of Earth's surface. Griggs's model used oil and sand and waterglass (sodium silicate) to get the effects he wanted. Here both drums are rotating, forming valleys. (Reprinted by permission of the American Journal of Science, from D. Griggs, A theory of mountain building, *American Journal of Science* 237 [1939]: 611–50, fig. 14, 642.)

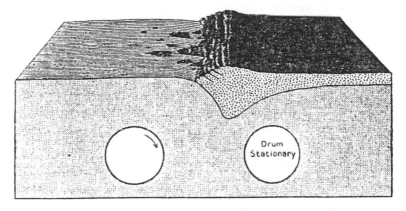

Figure 11. Just one drum rotates, and we get mountain building. (Reprinted by permission of the American Journal of Science, from D. Griggs, A theory of mountain building, *American Journal of Science* 237 [1939]: 611–50, fig. 15, 643.)

THE MOUNTAIN BUILDING CYCLE

Geosyncline

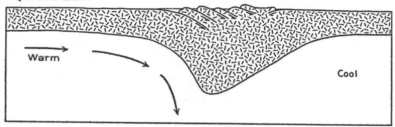

1. First stage in convection cycle — Period of slowly accelerating currents.

Region of Tension

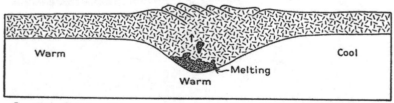

2. Period of fastest currents — Folding of geosynclinal region and formation of the mountain root.

3. End of convection current cycle — Period of emergence. Buoyant rise of thickened crust aided by melting of mountain root.

Figure 12. The mountain-building cycle. Note how it depends on the variable rate of circulation explained by the model in figure 10. (Reprinted by permission of the American Journal of Science, from D. Griggs, A theory of mountain building, *American Journal of Science* 237 [1939]: 611–50, fig. 16, 644.)

Griggs's dual modes of mechanistic thinking are clear. Obviously, he relied on the laws of physics and chemistry to show how rocks rise because of heat and sink because of gravity, how fluids flow and at what speeds, how materials of different densities find their respective levels of equilibrium (Archimedes' principle), and much more. But, equally obviously, he was thinking in terms of models, meaning mechanical contraptions. This is true of his model of convection and how it works in cycles; it is also true of his model of rotating fluids, which shows how they can bring on thickening of the crust and what happens subsequently. In other words, Griggs was thinking of the geology of the world as if it were an artifact, as if it had been designed by someone and then built, and, as the figures show, he actually built scale models to prove his points. The same kind of thinking is evident as we come to modern geology, the theory of continental drift driven by plate tectonics (Keary, Klepeis, and Vine 2009). Scientifically, Griggs has taken us halfway there. Subterranean convection currents do exist, carrying along great chunks—"plates"—of the crust, and on these ride the continents (fig. 13). The plates arise from the depths—for instance, from the center of the Atlantic ocean—and move very slowly around the globe. When they meet, one slides down beneath the other until, deep enough down, it liquefies, and the convection current keeps going and recycling (figs. 14 and 15).

What is important for our purposes is that this theory is mechanical in the sense of seeing the world as governed by natural law—blind regularities that hold unceasingly without exception. It is mechanical also in the sense of seeing Earth as a kind of machine, with subterranean fluids cycling endlessly, bringing about the surface phenomena of the globe on which we all live. And note that, as for Hutton, this is not any old machine, but a machine of the Industrial Revolution that goes right back to the Newcomen engine. Central heat drives everything, with liquids and gases being formed and forcing movements, and feedback occurs at many points as one part heats and sets off reactions that cause other parts to compensate. Most importantly, this creates some kind of

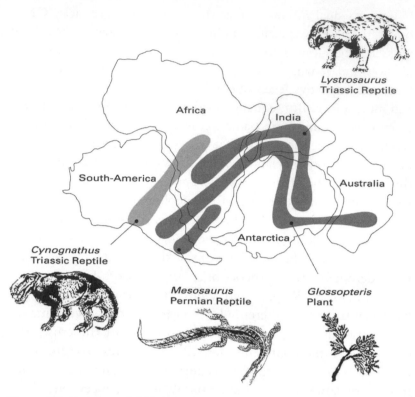

Figure 13. Continental drift, inferred from (and explaining) the distribution of fossil animals and plants. (From the United States Geological Survey website: http://pubs.usgs.gov/gip/dynamic/continents.html [accessed 12/12/12].)

balance or steady-state system, for as one plate sinks, another rises. Yet this is a machine like Hutton's with no ultimate purpose or reason. As far as science is concerned, it is not a machine made by God, and it is certainly not a machine made with humans in mind. In this significant sense it is entirely blind. We do not ask, "What is the point of continental drift?" It just is. In other words, the machines that Griggs was building and the machine represented by plate tectonics are akin to the contraptions you might find in the contemporary wing of the Art Institute in Chicago or the Tate

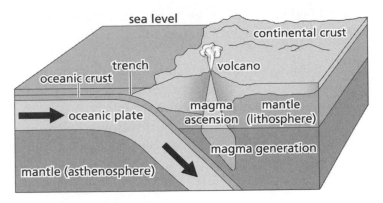

Figure 14. Plate tectonics, showing how, as the plates revolve and slip beneath the surface, we get mountain formation, volcanoes, and earthquakes. (Drawing courtesy of Martin Young.)

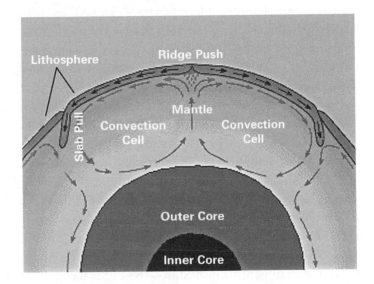

Figure 15. An overview of the workings of plate tectonics, showing how heat keeps the system in constant motion. (Drawing courtesy of Martin Young.)

Modern in London, which just go round and around. There is no purpose to their motion; there is no teleology. There is no intrinsic value.

In a way, this is simply in agreement with the thinking of Robert Boyle. In the physical world, final causes are no longer part of the conversation. And this points us toward biology, because it was there that, from Boyle through Kant, final causes were making their last stand. In our two-hundred-year period, what happened in that science? Could mechanism, in one or both senses, make inroads there?

DARWINIAN BIOLOGY

Since final cause is the object of our inquiry, let us jump straight to the man who is generally considered to have cracked that problem, Charles Lyell's fellow uniformitarian, Charles Darwin (1809–82), whose *Origin of Species* was published in 1859. Darwin was not the first evolutionist. Among others, his grandfather (Erasmus Darwin) arrived at the idea before him. What Darwin did was make the idea of evolution commonsensical (Ruse 1999a, 2008). Part of his success lay in his attack on the problem of final cause—in terms of the hand and the eye. He set about explaining them in a purely mechanistic fashion. Darwin was very much aware of final cause (this was the language that he himself used) and thought it was possibly the most significant feature of the organic world. He got this from theologians, scientists, and philosophers. As a young man at the University of Cambridge, he had read with much care the works of Archdeacon William Paley, including his famous exposition of the argument from design—the eye is like a telescope; telescopes have designers; therefore the eye must have a designer, none other than the divine ophthalmologist. In his *Autobiography* (1969), Darwin joked that he could have written Paley out by heart—but he wasn't joking.

But how do we account for final causes in biology using an entirely mechanistic strategy (in the sense of blind law and nothing else)? Enter Darwin's great discovery: natural selection. Coming from rural England, Darwin was fully aware of the power of artificial selection in the hands of a skillful breeder (Browne 1995). One can change animals and plants almost indefinitely—fatter pigs, beefier cows, shaggier sheep, bigger turnips, fancier pigeons, fiercer dogs. By noting desirable features and using their possessors as the progenitors of the next generations, breeders can achieve outstanding results. "Lord Somerville, speaking of what breeders have done for sheep, says:—'It would seem as if they had chalked out upon a wall a form perfect in itself, and then had given it existence.' That most skilful breeder, Sir John Sebright, used to say, with respect to pigeons, that he 'would produce any given feather in three years, but it would take him six years to obtain head and beak'" (Darwin 1859, 31).

The big question was how to arrive at a natural form of such a process? Darwin had read the *Essay on a Principle of Population* (the sixth edition of 1826) by the Reverend Thomas Robert Malthus. Concerned about false notions of social progress, Malthus argued that population numbers will always tend to outstrip supplies of food and space, resulting inevitably in a struggle for existence. Darwin seized on this idea and envisioned a struggle for life and reproduction throughout the whole animal and plant world. "Although some species may be now increasing, more or less rapidly, in numbers, all cannot do so, for the world would not hold them" (Darwin 1859, 63–64). From this, given the huge amount of variation that exists among organisms, even those of the same species, natural selection followed readily. Can it be "thought improbable, seeing that variations useful to man have undoubtedly occurred, that other variations useful in some way to each being in the great and complex battle of life, should sometimes occur in the course of thousands of generations?" Can it be doubted "that individuals having any advantage, however slight, over others, would have the best chance of surviving and of procreating their kind? On the other hand, we may feel sure that any variation in the least degree injurious would be rigidly destroyed. This preservation of favourable variations and the rejection of injurious variations, I call Natural Selection" (80–81).

The most significant aspect of natural selection (as noted in chapter 1) is that it is a process for change of a particular kind, namely, in the direction of (what biologists call) adaptive advantage. It produces things like the hand and the eye. In other words, the idea addresses the issue of final cause. It provides a naturalistic explanation of the end-directed nature of organic features or characteristics. But note that it does so by setting living things in the domain of law that is in no sense guided—that is blind, as it were. Darwin was explicit in this intention. "The lightning kills a man, whether a good one or bad one, owing to the excessively complex action of natural laws,—a child (who may turn out an idiot) is born by action of even more complex laws,—and I can see no reason, why a man, or other animal, may not have been aboriginally produced by other laws" (Letter to Asa Gray, May 22, 1860, in Darwin 1985, 8:224). There is no teleology here—either in Plato's sense (conscious intention) or in Aristotle's (special vital forces doing their work).

MECHANISM?

Mechanism in the first sense, that of constant law, was always the Darwinian game plan. Rather cheekily, given that his friend and mentor, the philosopher William Whewell, thought that organic origins were beyond science and demanded miracles, Darwin put opposite his title page for the *Origin of Species* the following passage by Whewell from a work on natural theology: "But with regard to the material world, we can at least go so far as this—we can perceive that events are brought about not by insulated interpositions of Divine power, exerted in each particular case, but by the establishment of general laws." What of the second sense of mechanism, meaning "machinelike"? For the moment, pass over natural selection itself, and focus on its products, namely, the adaptive features of organisms. Darwin offers much discussion in the explicit language of this sense of mechanism. This is not surprising, because this was the language of the natural theologians: "Movable joints, I think, compose the curiosity of bones; but their union, even where no motion is intended or wanted, carries marks of mechanism and of mechanical wisdom" (Paley 1819, 86). Darwin echoed this idea, only dropping the theological implications: "Almost every part of every organic being is so beautifully related to its complex conditions of life that it seems as improbable that any part should have been suddenly produced perfect, as that a complex *machine* should have been invented by man in a perfect state" (Darwin 1959, 121, from 1861 edition; my italics).

There is obviously a difference between geology and (Darwinian) biology. In geology there simply isn't any end, even for the most machinelike system. A mountain is a mountain is a mountain. In biology, there are ends for the machines. A "fertilization machine" in a plant serves the end of reproduction. Speaking of final causes is thus legitimate in the biological context. Metaphysically speaking, however, the difference is not that profound. Once you have related something to the well-being of its possessor then, as in geology, you come to a dead end. In the cosmic scheme of things, mountain building serves no purpose. In the cosmic scheme of things, the plant's reproduction serves no purpose. It is true that a cow might be interested in the reproduction of the plant, but the plant does not reproduce in order to feed the cow. Admittedly, when you have a symbiotic relationship like that between flower and insect, one can speak thus in a limited way. The flower makes a sweet substance to attract the bee, and so forth. The bee would not be attracted unless there was a sweet substance. So organic machines are not quite purposeless in the sense that we find in geology, at least not in the immediate sense. But ultimately bee and plant serve no ends—biologically speaking, that is. We like honey, and we like pretty flowers, but those are our ends, not nature's.

REDUCTION

Switch the focus now from mechanism to the related notion of reduction. In the case of geology, there is no real argument or major point to be made. Obviously, the triumphs of mechanism are at the same time triumphs of reductionism—the world is explained

in terms of its parts. Evolutionary biology is rather more complex
and interesting. At one level, the story is as straightforward as in
geology. If you look at the history of the subject since Darwin, a
tendency to reduction has been the dominant theme. Darwin had
no real idea of heredity and its causes, although in the 1860s he did
hypothesize little particles, "gemmules," that might be the causes
behind the transmission of characteristics from one generation
to the next (Olby 1963). The big breakthrough came from Darwin's contemporary, the obscure Moravian monk, Gregor Mendel
(Bowler 1989). When his ideas were rediscovered at the beginning
of the last century, they were quickly developed and extended into
the "classical theory of the gene," according to which the units
of heredity were carried on the chromosomes within every cell
of an organism's body. Mendelian genetics and Darwinian theory
were soon combined to make modern evolutionary theory, which
became an even stronger testament to the worth of reduction as
molecular biology developed and also incorporated evolutionary
thought (Ruse 1999b). All of this has led to a reconceptualizing
of evolutionary thinking, with the gene now often taken to be the
focus of selection. Combining this reductionism with the second
sense of the mechanistic metaphor, Richard Dawkins describes
things thus in the *Selfish Gene*: "We are survival machines, but 'we'
does not mean just people. It embraces all animals, plants, bacteria, and viruses." He goes on to say, "Different sorts of survival
machines appear very varied on the outside and in their internal
organs. An octopus is nothing like a mouse, and both are quite
different from an oak tree. Yet in their fundamental chemistry they
are rather uniform, and, in particular, the replicators which they
bear, the genes, are basically the same kind of molecule in all of
us—from bacteria to elephants." This leads to his conclusion: "We
are all survival machines for the same kind of replicator—molecules
called DNA—but there are many different ways of making a living
in the world, and the replicators have built a vast range of machines
to exploit them. A monkey is a machine which preserves genes up
trees, a fish is a machine which preserves genes in the water; there

is even a small worm which preserves genes in German beer mats. DNA works in mysterious ways" (Dawkins 1976, 22).

There is another level to the reductionism story that is pertinent to our analysis. Remember that Darwin's theory did not come out of nowhere. It was very much a product of his background—his family and his training. We have seen evidence of this already in the extent to which his reading of Paley (part of his course as an undergraduate at Cambridge) influenced his thinking on final causes. His family was even more important. The Darwins and the Wedgwoods—the family of his mother and of his wife—were absolutely children of the Industrial Revolution. Paternal grandfather Erasmus was a close friend of Boulton and Watt and was himself a great inventor of machinery. Maternal grandfather Josiah Wedgwood (also a friend of Erasmus Darwin) was one of the great leaders of the Industrial Revolution, being responsible for the total transformation of the pottery trade. The Darwin-Wedgwood family accepted completely the virtues of the Industrial Revolution—they made a great deal of money out of it—and this included the philosophy that lay behind and drove the revolution: the liberal economic philosophy developed and promulgated by the political economists of the eighteenth century, especially by Adam Smith. Much of Smith's thinking finds its way virtually unchanged into Darwin's work. The division of labor is a metaphor to which Darwin returns again and again—at the individual level in the body and at the group level in populations or societies. In the same vein, we have seen already the enthusiasm with which Darwin embraced Malthus's idea that population pressures lead to a struggle for existence. As with natural theology, Darwin turned Malthus on his head. The political economist used the idea of the struggle to suggest that state welfare was harmful because it only exacerbated the problem. The biologist used the struggle to drive the process of ongoing change. But this was in biology. In real life, Darwin was part of the capitalist system. He wanted hungry workers to labor in the Wedgwood factories and was very much against attempts to alleviate the struggle formally; for example, he had little time for unions.

Return now for a moment to the notion of mechanism in the second sense of actual machines. As students of the eighteenth century have pointed out, a strong connection exists between the theorizing of the economists and the technology of the age (Mayr 1989). In both cases things work according to unguided law—although admittedly both made the deistic assumption that God lay behind the laws—and, moreover, things work in a way that involves feedback, fueled by fire in the one case and greed or self-interest in the other. This is all bound up with the kind of representative democracy that Britain was embracing, which had developed after the deposing of the authoritarian Catholic King James the Second. Both economics and technology included a dimension of liberty, expecting and allowing space for changes in response to events. That is what feedback is about. The connection is more striking when we consider the Continent at the same time, where the older metaphor still predominated—everything going according to clockwork!—and where societies were dominated by powerful rulers who were not about to allow the dimensions of freedom demanded by the Industrial Revolution.

The pertinence of all of this to Darwinian selection is immediate and obvious. In natural selection itself, Darwin did have a mechanism in the second sense, although it was not modeled directly on technology—the Newcomen engine—but on economics. Note that this is not an "either-or" situation. With their joint highlighting of feedback, the technology and the economics echo each other, so in a sense selection too is echoing the machine successes of the Industrial Revolution. If we focus on economics rather than technology, we see the pertinence of this discussion for reductionism. We think now of social units rather than physical entities, and, in particular, we think of individuals rather than groups, for the liberal economics of the eighteenth century put everything firmly on the individual. Tradesmen are not uniting to look after themselves as a group; they are acting in their own best interest as individuals, and any good that comes out of this—for them or for me—is a result of this thinking. "It is not from the benevolence of the butcher, the brewer, or the baker that we expect our dinner, but from their regard to their own interest" (Smith 1976, sec. 2A, 26–27). In other words, at the natural level, explanation is from the bottom up rather than from the top down. Charles Darwin completely accepted this mode of thought. For him, the basic unit in the evolutionary process is always the individual rather than the group. The struggle is between one organism and another, or one organism for itself, and never between or for groups. He writes, "Hence, as more individuals are produced than can possibly survive, there must in every case be a struggle for existence, either one individual with another of the same species, or with the individuals of distinct species, or with the physical conditions of life" and goes on to say, "It is the doctrine of Malthus applied with manifold force to the whole animal and vegetable kingdoms" (Darwin 1859, 63).

Darwin may at first have slipped into this mode of thinking without much reflection. It came with the territory, as it were. But it was not long before he found himself having to think through and justify this kind of reductionism about individuals—or, to use the language introduced in chapter 2, to explain why he was an individual selectionist rather than a group selectionist. (This was never Darwin's language.) Above all, there was the problem of the social insects. If everything is for the individual—and remember, Darwin lacked an adequate theory of heredity, so he was thinking at the level of the organism and not, like Richard Dawkins, at the even more reductionistic level of "selfish genes"—then how can we explain what goes on among the ants and the bees, where the workers are sterile and devote all of their efforts to the good of the nest? Darwin's solution was to argue that the family can act as a unit, as an individual, so that workers are viewed as parts of the whole just as physical features such as eyes and teeth are parts of the whole. Hence, the nest as a whole is reproducing. Even in the case of human beings, where one might think that the development of conscience and a moral sense would lift people from selfish regard for themselves, Darwin refused to budge. Either help given to others is a function of what today is known as "reciprocal altruism"—you scratch my back and I'll scratch yours—or it is something that takes place within the tribe, and Darwin made it clear in an unpublished letter to his son George (April 27, 1876) that he regarded the tribe as akin to a nest of bees or wasps. Socially, Darwin would have seen this extension of the individual to the family as no concession at all. Darwin's biography reveals the extent to which he was embedded in family. Being often sick, staying at home tended by his wife (who was also his first cousin, with whom marriage was not quite dictated but strongly encouraged), surrounded by many children, visiting relatives (his older sister was married to his wife's brother) for holidays and relaxation, even staying with older brother Erasmus on visits to his dentist in London, he always thought in terms of the family versus others. The Darwin-Wedgwood clan could have given lessons to the Corleones.

Do not underestimate the significance or power of Darwin's thinking on what we would call today the "levels of selection problem." As always—Malthus provided the paradigm—although Darwin used the ideas of earlier thinkers, he transformed these ideas significantly as he set them in his own context. This comes through dramatically when we look, in the light of his commitment to individual selection, at Darwin's thinking on questions of balance and equilibrium. Adam Smith and his fellows, working within a British natural theological context, thought that self-interest leads not only to division but also to balance. Thanks to the "invisible hand" of God, unchecked society is in equilibrium, working with full efficiency. This societal equilibrium reflects what in the organic world is called the "balance of nature"—what Linnaeus at that time was calling the "economy of nature." It is obviously a notion that goes back to Plato and was to be welcomed by Romantics such as Schelling. The world is an integrated organism, and everything is in harmony, in balance. As one species grows, another arises to check the growth and is checked in turn by other factors, organic or inorganic.

Darwin had no objection to balance or equilibrium as such. In the case of individual organisms, it is or can be an adaptation like any other, brought on by selection for the benefit of the individual.

Temperature regulation in mammals serves the good of the individual or its family as much as fur and teats. He also saw it exhibited in larger entities that we might call "ecosystems." But he did not assert a natural tendency to balance. In an unpublished earlier version of the *Origin*, written in the mid-1850s, Darwin made this point explicitly. On introducing the struggle for existence, and having made the point that the struggle does not necessarily mean actual physical fighting, he added, "In many of these cases, the term used by Sir C. Lyell of 'equilibrium in the number of species' is the more correct, but to my mind it expresses far too much quiescence. Hence I shall employ the word struggle" (Darwin 1975, 187). In the published version of the *Origin*, Darwin followed up on this insight: "Battle within battle must ever be recurring with varying success; and yet in the long-run the forces are so nicely balanced, that the face of nature remains uniform for long periods of time, though assuredly the merest trifle would often give the victory to one organic being over another" (Darwin 1859, 78). Balance has to be of adaptive worth to individuals—all sharing in a harmonious existence. Sometimes this holds; sometimes not. Apart from the fact that, as an evolutionist, Darwin saw everything tending to change (that is, to break from equilibrium), group balance or equilibrium is not in itself of biological worth. It persists as long as it is needed by individuals, and not a moment longer.

EVOLUTIONARY BIOLOGY TODAY

In the hands of Charles Darwin, mechanism comes to biology. His theory is mechanistic in the first sense of working according to blind law and also in the second sense, both for selection itself as modeled on the mechanisms of economics and for the adaptive features produced by selection as functioning pieces of technology, serving the needs of their possessors but ultimately having no overall purpose, any more than one finds in geology or any other physical science. Significantly, however, Darwin also showed why final cause had such a hold on biology. Thanks to natural selection, organisms—and their parts—appear as if designed, as if planned, so that there is a hint of teleology. Indeed, such thinking and language are permissible, and (as Kant argued strongly) heuristically invaluable. If you want to understand the reason for a part of an organism, ask about its purpose. The important thing is that this kind of teleology implies no external designer in the Platonic sense and no internal special forces in the Aristotelian sense. It is purely a mechanism. In the Darwinian case, it is also reductionistic. Darwin clearly believes that knowing about smaller physical entities is important for knowing about wholes—though he himself made no great contribution in this direction—and in the sense that organisms in competition and conflict, which ultimately is what Darwinian evolution is all about, are considered from the perspective of the individual and never the group. Small really is beautiful.

Now, as in the case of geology, let us jump to the present. Blind law rules supreme. Evolutionary biology is thoroughly mechanistic in this sense. Nor do Darwinians see any ultimate purpose in anything. Life just is, and that is all there is to it. Darwinian evolutionary biology is thoroughly mechanistic in the second sense also. Thinking in terms of machines is rife. That is how you do the science. Consider the strange plates along the back of the dinosaur Stegosaurus. Ideas about their purpose are all technology-based. Perhaps they are weapons of attack or defense, used in fighting just as is military hardware. It seems that this is not the case, because studies show that they would snap off far too readily. Perhaps they are used in attracting mates, rather like billboards advertising one's wares? It seems that this is not the case, if only because both sexes have the plates, and females generally have no need of such advertising. Perhaps they are used for heating and cooling. The herbivorous brute would generate a huge amount of internal heat as its food was being digested and probably fermenting. It would need to get rid of this unwanted warmth. Conversely, the cold-blooded animal could use a shot of sunshine in the morning to get it going quickly. The plates look just like the plates used in the cooling towers at electrical plants for heat transmission. This now is the popularly accepted explanation (Farlow, Thompson, and Rosner 1976). Type-two mechanism down the line.

What about reduction? In this molecular age, especially in the area of heredity, reductionism obviously becomes more and more intense. That is just a fact of modern science, and Darwinians would be unapologetic about it. Questions of reductionism with regard to the levels of selection—individual versus group selection—are, as we know from chapter 2, considerably more controversial. In this chapter, let us simply note that those who would unambiguously identify themselves as Darwinian tend to be strongly inclined to the individual-selection end of the scale (Ruse 2006). After Darwin himself, the most significant contributor to this view was the Englishman William D. Hamilton (1964). Agreeing that selection normally works only for the individual, he devised what was later labeled "kin selection" to explain social behavior, as exhibited, for example, by the Hymenoptera (the ants, the bees, and the wasps). Simply, since relatives share heredity (genes), inasmuch as relatives reproduce, each one is doing so itself by proxy, as it were. Hamilton started with Darwin, but his knowledge of modern genetics allowed him to go further, arguing that we can now see how individual interests rule supreme even within nests of social insects. Following Hamilton came the American evolutionist George Williams (1966), a withering critic of most group-selection arguments, and the English evolutionist John Maynard Smith (1982), who, though he thought that sex maintenance might be the exception that proves the rule, was fanatical in his individualism as he applied game theory to problems of social behavior. Correspondingly, there was a flood of empirical work, on subjects from dung flies to red deer, showing how an individual-selection perspective leads to fruitful and confirmable predictions. But, as also noted in chapter 2, while this may be the majority position, not all would agree. We turn next to the voices of dissent.

5

ORGANICISM

William Wordsworth wrote "Tintern Abbey" in 1798. As much as Goethe, he showed thoughts of unity, with Earth integrated by world spirits, and especially the sense that in some way nature itself is vibrant, is living. Romanticism crossed the Atlantic too, and by the third and fourth decades of the nineteenth century, it was especially evident in New England, in the thinking of the so-called transcendentalists. Of the sometime Unitarian minister Ralph Waldo Emerson, the celebrated philosopher Charles Sanders Peirce was later to write, "I was born and reared in the neighborhood of Concord, at the time [1839] when Emerson and . . . friends were disseminating the ideas that they had caught from Schelling, and Schelling from Plotinus" (Peirce 1934, 6:102). The major linking influence was the English poet and philosopher Samuel Taylor Coleridge, who (thanks to support from Darwin's uncle and future father-in-law, Josiah Wedgwood the younger) introduced Schelling to the English-speaking world. Emerson expressed this vision in his celebrated essay "Nature," written in 1836: "Man is conscious of a universal soul within or behind his individual life, wherein, as in a firmament, the natures of Justice, Truth, Love, Freedom, arise and shine. This universal soul, he calls Reason: it is not mine, or thine, or his, but we are its; we are its property and men." This is not just a quality of humans: "The blue sky in which the private earth is buried, the sky with its eternal calm, and full of everlasting orbs, is the type of Reason. That which, intellectually considered, we call Reason, considered in relation to nature, we call Spirit. Spirit is the Creator. Spirit hath life in itself" (Emerson 1836, chap. 4, sec. 2). As we might expect, Platonism was a major theme. "Herein is especially apprehended the unity of Nature,— the unity in variety,—which meets us everywhere. All the endless variety of things makes an identical impression. . . . A leaf, a drop, a crystal, a moment of time is related to the whole, and partakes of the perfection of the whole. Each particle is a microcosm, and faithfully renders the likeness of the world" (chap. 5, sec. 2).

All of this seems dreadfully out of step with the actual science of the nineteenth century—not just in physics and chemistry, but in biology and geology also. Darwin's theory in the *Origin* is thoroughly mechanistic and reductionistic, and leads to even more thinking in that direction. Can one doubt that the spirit of Romanticism was brushed aside? I take up this question in this chapter and the next, concentrating here on those who would have repudiated indignantly any suggestion that they were straying beyond the bounds of science and looking later at those who had other aims than producing conventional science. All of the critics of mechanism share an underlying moral concern. This does not imply that mechanists are immoral people; it is just that they want to keep their science and their morality separate. The world of science is the world of meaningless matter, endlessly moving. That is all there is to it. In the words of Richard Dawkins, "The universe we observe has precisely the properties we should expect if there is, at bottom, no design, no purpose, no evil and no good, nothing but blind, pitiless indifference" (Dawkins 1995, 133). The critics want none of this. Although they are willing in varying degrees to mix their moral sentiments with their factual claims, their senses of value, of right and wrong, are clearly visible in their science. This is to be expected if the Platonic tradition is to carry any weight. The world was created according to the Form of the Good, and it reflects this fact. As we shall see, by and large, the professionalism of the scientists discussed in this chapter led them to keep their explicit values somewhat suppressed in their science. Nevertheless, they make it intentionally clear that their science is molded and directed by values. For them, science is a deeply moral enterprise, and in their hearts they see the world about us as something of value in its own right.

I use the term *organicism* for the philosophy I am describing here because of the historical continuity and because of the emphasis on integration, as one finds in the individual organism. Another term often used is *holism*, a notion invented by the early twentieth-century South African statesman, Jan Smuts (1926). Another term is *emergentism*, implying (in Aristotle's words) that the whole is more than the sum of the parts (*Metaphysics* 1045a10). These terms must be used carefully, for they can cover a range of rather different positions. Indeed, especially if what is being urged is the directive to consider all of the evidence, it is hard to see what distinguishes an organicist from a mechanist. Think of a clock, the paradigmatic example of a machine, which is surely more than the sum of its parts. It is not just bits and pieces of metal and glass, but something put together, organized, to tell the time—an emergent if ever there was one. Interestingly, but not entirely surprisingly, before the Scientific Revolution the clock was taken as a metaphor for the soul, for the vital functioning of consciousness (Neuman 2010).

Yet something important is at stake here that becomes evident if we adopt the heuristic principle that often it is easier to see what people do believe by focusing on what they do not believe. In the biological world, one sees both the moral thrust and the emphasis on unity particularly clearly in the attitude toward Darwinism, especially the mechanism of natural selection. No one denies evolution. That is a given. But the competition at the heart of Darwin's vision of change is considered deeply upsetting and in some respects offensive. Of course there is death and destruction, but it is ameliorated by the essential oneness of everything, the push for harmony and cooperation. Adaptation is not denied, but it is often downplayed, especially as a driving force for evolution. Darwin's focus on the individual is thought to be quite incomplete and misleading. If there is selection, then often (perhaps primarily) it is seen as promoting unity in the group, at some level or another. The notion of organism applies to groups as well as to individuals.

What about the question of teleology, of final cause (in the strong Platonic or Aristotelian version)? Whatever their dissatisfaction with naked mechanism, the scientists discussed in this chapter are generally children of the Scientific Revolution. There is no place in their science for conscious intention, such as we find in the thinking of today's so-called Intelligent Design theorists or of the closely related theistic evolutionists, who put God's intentions into the variations of evolution and so forth (Pennock and Ruse 2008). Even those who are Christians—some are, some aren't—don't want to go there. Nor do they want to suppose nonconscious, Aristotelian forces directing things. At the beginning of the twentieth century, there was a group of "vitalists," led by the German embryologist Hans Driesch (1908), who called such forces "entelechies," and the French philosopher Henri Bergson (1907), who called such forces "*élans vitaux.*" The organicists pride themselves on steering between mechanism and vitalism. Still, their emphasis on the concept of the group does allow for some of the benefits of vitalism that mechanism does not, a kind of faux teleology, if you will, where an end harmony emerges naturally. In a way that makes a pure Darwinian uncomfortable, they are looking for balance, for integration, for equilibrium. They seek value. They are after an attenuated Platonic teleology, brought on by attenuated Aristotelian forces. Let us use the focus on the group, on the whole, on balanced harmony as our thread of Ariadne to guide us through this chapter.

HARVARD HOLISM

We start by introducing one of the more interesting figures in our story, the mid-nineteenth-century Englishman Herbert Spencer, long belittled and despised but now seen as having left his mark on much twentieth-century biology, particularly twentieth-century American biology in the areas of evolution and ecology.

In a discussion of organicism, Spencer might seem to be quite out of place. He is best known today as the quintessential social Darwinian, promoting an extreme form of laissez-faire socioeconomics, the epitome of someone who sees bloody struggle at all levels, among humans and all other organisms. Such a misconception about Spencer is only equaled by the denial of his lasting influence (Richards 1987). It is true that he believed in struggle and laissez-faire and, like Darwin, he had been deeply influenced by Malthus, but his reading of these ideas was completely different from that of Darwin. Although he discovered natural selection independently of Darwin and also coined the alternative phrase "survival of the fittest," Spencer was never very keen on selection as a mechanism. He was always primarily a Lamarckian, believing in the inheritance of acquired characteristics. For Spencer, the struggle was less a matter of winners and losers, and more about a process that led all sides to develop new and admirable features that would then be passed on. He combined this with the Victorian belief that we have only so much vital bodily fluid and that its use in reproduction prevents its use in powering brains. Thanks to struggle and consequent development, however, organisms progress upward to incorporate larger thinking organs and with this comes a consequent decline in reproduction and subsequent minimizing of the struggle. For Darwin, we will always have Malthus with us. For Spencer, in the end Malthus will become irrelevant (Young 1985).

Spencer thought that systems, which include biological systems such as a species or a group of species, normally exist in equilibrium, being at first relatively homogeneous. Then something disturbs the stability, and change occurs. The system is always striving to regain stability, however, and eventually does so, but at a higher level than previously, meaning at a level with greater heterogeneity—with greater intrinsic value (Spencer 1862). And running through this vision—known as *dynamic equilibrium*—was an ardent holism that saw a kind of organic unity existing at all levels and across all types or systems. In a celebrated essay in which he likened a society to an organism, Spencer highlighted four features linking individuals and groups: first, they grow from very small units to much larger ones; second, they become increasingly complex as they grow; third, this complexity increasingly involves a far greater mutual dependence of the parts on each other (underlying this would be Adam Smith's division of labor); and fourth, "the life of a society is independent of, and far more prolonged than, the lives of any of its component units" (Spencer 1868, 392; originally published in 1860). As each individual's parts undergo change in growth, maturity, and aging, so in society one finds analogous change and continuity.

Spencer was rarely generous in acknowledging influences on his thinking. Obviously, both in timing and in spirit, he is completely non-Darwinian. For Darwin, equilibrium is an uncomfortable notion. For Spencer, equilibrium is central. More positively, we can be certain of the importance of Schelling in Spencer's thinking, as mediated by Coleridge: "I should add that the acquaintance that I accidentally made with Coleridge's essay on the Idea of Life, in which he set forth, as though it were his own, the notion of Schelling, that Life is the tendency to individuation, had considerable effect" (Duncan 1908, 541). Anticipating thoughts of progress from the homogeneous to the heterogeneous, Spencer would have learned from the poet that "the living power will be most intense in that individual which, as a whole, has the greatest number of integral parts presupposed in it," and that by "Life I everywhere mean the true Idea of Life, or that most general form under which Life manifests itself to us, which includes all its other forms. This I have stated to be the *tendency to individuation*, and the degrees or intensities of Life to consist in the progressive realization of this tendency" (Coleridge 1848, 49). Unambiguously, there is the *Naturphilosoph*'s developmental self-organization, combined with a holistic perspective on every aspect of existence—something Spencer dressed up in the language of British social theory.

That Spencer should in turn have influenced American biology is not surprising. On the one hand, he enjoyed worldwide popularity toward the end of the nineteenth century, especially in America, where he had enthusiastic supporters at all levels of soci-

ety. People liked the idea of ever-improving development, progress, that was built into the Spencerian system. On the other hand,
there was a gap waiting to be filled because, although Darwin had
converted people to evolution, there was little enthusiasm for natural selection. Neither is it surprising that Spencer's ideas went
right to the top. After all, Harvard University is in Cambridge,
Massachusetts, the heart of American transcendentalism (Menand
2001). In addition, Harvard is where the Swiss-born ichthyolo-
gist Louis Agassiz made his home after crossing the Atlantic in
1846 and developed his friendships, especially with Emerson. The
European scientist's thinking was deeply infused with the spirit of
Romanticism: as a young man, he had gone to Munich to learn
at the feet of none other than Friedrich Schelling. In the words of
Agassiz's fellow student (and future brother-in-law), the botanist
Alexander Braun, "A man can hardly hear twice in his life a course
of lectures so powerful as those that Schelling is now giving on
the philosophy of revelation" (Agassiz 1885,1:91). Agassiz today
is best known as the anti-evolutionary opponent of the botanist
Asa Gray, a champion of Charles Darwin. But do not be misled. It
was Agassiz who had the influence and the students, all of whom
imbibed his holistic philosophy especially the emphasis on form
over function and on Goethe-like archetypes rather than Paley-like
adaptations—and molded it to evolutionary ends (Ruse 1996).

Spencer's holistic vision was taken up with enthusiasm by an
influential group of life scientists. A prominent member was the
biochemist L. J. Henderson who, having started work in the first
decade of the new century on issues to do with equilibrium in
the human body, was soon led to broader interests and questions
(Parascandola 1971), eventually making his famous claim that not
only are organisms adapted to their environment, but in some
sense the environment is adapted to the organisms. Anticipating
what today is known as the "anthropic principle," Henderson
looked at a number of common substances that are essential to
the functioning of organisms—carbon, hydrogen, and oxygen—
stressing how truly remarkable it is that these substances have or
give rise to the properties that they do. For instance, water—the

combination of oxygen and hydrogen—"possesses certain nearly unique qualifications which are largely responsible for making the Earth habitable, or at least very favourable as a habitation for living organisms" (Henderson 1913, 85). Crucially, it has a high heat capacity, meaning that the temperatures of oceans and inland waters remain fairly constant, and thus moderate Earth's temperature in both winter and summer. This is also of vital importance in keeping human bodies at a fairly constant temperature, especially during strenuous work.

There is a level of integration here that certainly seems to foreshadow the Gaia hypothesis, although I am not sure that Henderson was really thinking this way. He seems to think of the inorganic world as having certain features necessary for the organic world without thinking of any causal connections. But Henderson was thinking in a holistic mode, seeing things as working for the good of the whole. Why should all of this be so? Henderson was convinced that mechanism played a role in understanding the world, including the organic world, but not an exclusive role. There does seem to be some kind of built-in design to the laws of nature. Henderson did not want to invoke God, and so left the existence of such design as a brute fact. He argued that "our new teleology cannot have originated in or through mechanism, but it is a necessary and preëstablished associate of mechanism. Matter and energy have an original property, assuredly not by chance, which organises the universe in space and time" (Henderson 1913, 308). He kept returning to this thinking, showing how impressed he was by the way that systems, from organisms on up, exhibit or return to equilibrium. "In short the primary constituents of the environment, water and carbonic acid, the very substances which are placed upon a planet's surface by the blind forces of cosmic evolution, serve with maximum efficiency to make stable, durable, and complex, both the living thing itself and the world around it" (Henderson 1917, 5). Naturally enough, Henderson praised Spencer for such thinking. He was more a visionary than an empirical scientist, but "his generalizations, regarded as provisional and tentative hypotheses, possess genuine importance" (Henderson 1917, 124).

Henderson made much of the fact that equilibrium was not exclusively a Spencerian notion but was shared by genuine empirical researchers. It was central to the thinking of the mid-nineteenth-century French physiologist Claude Bernard. Bernard's notion of a *milieu intérieur*—an internal stable system in the organism—was accepted enthusiastically, although Henderson pointed out that Herbert Spencer's characterization of life as adjustment of the internal to the external "is really a true statement of the phenomena of organization" (Henderson 1917, 83). His enthusiasm for Bernard's work was shared by Henderson's long-time colleague, the physiologist Walter B. Cannon (Benison, Barger, and Wolfe 1987). In a popular book whose title, *The Wisdom of the Body* (1931), revealed his quasi-teleological sympathies, Cannon spoke of equilibrium in language seized on by James Lovelock: "The coordinated physiological processes which maintain most of the steady states in the organism are so complex and so peculiar to living beings—involving, as they may, the brain and nerves, the heart, lungs, kidneys and spleen, all working cooperatively—that I have suggested a special designation for the states, *homeostasis*" (24).

Like Henderson, Cannon was a serious scientist whose thinking was based firmly on major empirical investigations, although they both were thinking in a broader fashion. Cannon stressed how homeostasis leads to a greater dimension of freedom for the individual. Because of it, we are able to "analyze experience, we move from place to place, we build airplanes and temples, we paint pictures and write poetry, or we carry on scientific researches and make inventions, we recognize and converse with friends, educate the young, express our sympathy, tell our love—indeed, by means of it we conduct ourselves as human beings" (Cannon 1931, 302–3). And, in the tradition of just about everyone, especially Spencer, Cannon concluded his discussion by drawing analogies between individuals and societies, showing that the moral realm was close to the surface in his science. Just as individuals have self-regulating abilities (less in simple organisms, more in complex organisms), so societies have self-regulating abilities (less in simple societies, more in complex societies). "At the outset it is noteworthy that the body politic itself exhibits some indications of crude automatic stabilizing processes" (311). Not that all is entirely well: "It would appear that civilized society has some of the requirements for achieving homeostasis, but that it lacks others, and because lacking them it suffers from serious and avoidable afflictions" (312–13).

A third influential figure in the Harvard group was the ant specialist William Morton Wheeler (Evans and Evans 1970). An ardent Spencerian, he endorsed entirely analogies between human and ant societies, going on to speak of ant societies as being akin to individual organisms. "An ant society, therefore, may be regarded as little more than an expanded family, the members of which cooperate for the purpose of still further expanding the family and detaching portions of itself [to] found other families of the same kind. There is thus a striking analogy, which has not escaped the philosophical biologist, between the ant colony and the cell colony which constitutes the body of a Metazoan animal." He added that the "queen mother of the ant colony displays the generalized potentialities of all the individuals, just as a Metazoan egg contains *in potentia* all the other cells of the body. And, continuing the analogy, we may say that since the different castes of the ant colony are morphologically specialized for the performance of different functions, they are truly comparable with the differentiated tissues of the Metazoan body" (Wheeler 1910, 7). The unit of evolution is therefore the group rather than the individual ant.

Wheeler was not overly enthusiastic about natural selection; along with Spencer, he emphasized cooperation as an important aspect of organic life as opposed to struggle. Ultimately, though, much is clouded in mystery. "We can only regard the organismal character of the colony as a whole as an expression of the fact that it is not equivalent to the sum of its individuals but represents a different and at present inexplicable 'emergent level'" (Wheeler 1928, 24). Like his colleagues, Wheeler was interested in the implications of all of this for human society, and (reflecting his own area of expertise) did not feel unduly optimistic. Given "the atrophy or subatrophy of our organs and tissues brought about by the ever-increasing specialisation in our activities," unfortunately, "we can hardly fail to suspect that the eventual state of human society may be somewhat like that of the social insects—a society of very low intelligence of the individuals combined with an intense and pugnacious solidarity of the whole" (Wheeler 1939, 162). Notice that, although Darwin felt compelled to treat the ant colony as an individual, Wheeler went way beyond Darwin in thinking of human society as akin to an ant colony. Wheeler was prepared to treat a group of unrelated humans as a unit of selection in a way that Darwin would never have done. For Darwin, it was the relatedness that counted; for Wheeler, it was the group.

ECOLOGY AT CHICAGO

If we use the term *ecology* here to mean the formal science of the interaction of organisms with their environment, and leave until later the more popular philosophy of nature that goes under that name, we can say that ecology began to develop in America in a serious way around the beginning of the twentieth century. Many of its early practitioners came not from the established academic institutions of the East, but from the newer universities in the Midwest and beyond—places where the interactions of land and organisms were not only easier to study but of vital agricultural and recreational significance. At the new university founded in Chicago at the end of the nineteenth century, ecology found a welcoming home. From its beginning, the university was strong in biology, and, thanks especially to its founding chair, Charles Otis Whitman, the underlying philosophy was the Spencerian, holistic perspective of the biologists at Harvard (Mitman 1992). For ecology, the key figure was Warder Clyde Allee, whose science was infused also by his lifelong membership in the Religious Society of Friends (the Quakers), a branch of Christianity in which moral imperatives equal or outweigh any of the more traditional theological concerns, and whose members play down the competitive nature of life to stress cooperation and integration. Allee, who graduated from the University of Chicago before becoming a faculty member, was trained as a physiologist. In Allee's early years (the 1920s), therefore, evolutionary questions were not at the forefront, nor indeed was genetics. When dealing with the reasons why organisms come together into groups ("aggregations"), he saw benefits as coming more from common cause than from shared lineages or from the integrating effects of breeding. Thus, in a major discussion in the *Quarterly Review of Biology*, he stressed that social factors were involved only in a negative sense: "The only social trait necessary in many cases is that the animals shall be willing to tolerate the close proximity of other individuals" (Allee 1927, 371). What matter most are the things that a group can do better than the isolated individual. Production of heat is one such thing, and this, for Allee, was a major reason why social insects like the ants or the bees stick together. Moisture regulation is another factor. Allee noted as a third reason "the possible protection from toxic substances furnished by a mass of animals as compared with that of a single individual exposed to the same intensity of the toxic agent" (380). But Allee avoided arguing that the group owed its origins to the family. Perhaps reflecting a worry that the notion of family implies unequal power relationships (i.e., male dominance), a notion totally repudiated by the equality of women and men in Quaker theology and social structure, Allee explicitly endorsed Spencerian suggestions over Darwinian, arguing that "colony life arose from the consociation of adult individuals for cooperative purposes" (391). He continued, "In terms of human society, this view would stress the importance of the gang, rather than the family, as a preliminary step in the evolution of the social habit. *It is important to note that the gang cuts across family lines in its formation*" (my emphasis, to emphasize how deeply non-Darwinian is this line of thinking).

In 1925 the Harvard-educated geneticist Sewall Wright joined the biology faculty at the University of Chicago, where he remained until forced into retirement in 1955. Although geneticists were still regarded with some suspicion, Wright was welcomed at

Chicago, and he felt at home. In the early 1930s Wright produced a highly influential model of evolutionary change—the so-called shifting balance theory of evolution (Wright 1931). He saw large populations of organisms as fragmenting into small groups. Within these groups, even if selection was operating, the chance factors in reproduction might overwhelm all other factors, and hence new features could be formed randomly, by "genetic drift." The small populations might then rejoin and, through a kind of higher level of selection, preserve and spread the better features. Using the innovative metaphor mentioned earlier, Wright (1932) argued that evolution operates as if taking place on a hilly landscape. Populations sit on the top (representing the fittest places to be), and then, through drift and other random factors, they move down and into the valleys before shooting up to higher peaks elsewhere.

If the underlying Spencerian themes are not evident from the notion of being in a stable state before disruption and movement to a higher point of stability, then they are in Wright's language. "Evolution as a process of cumulative change depends on a proper balance of the conditions, which, at each level of organization— gene, chromosome, cell, individual, local race—make for genetic homogeneity or genetic heterogeneity of the species. . . . The type and rate of evolution in such a system depend on the balance among the evolutionary pressures considered here" (Wright 1931, 158).

Even if you were unaware that Wright subscribed to a philosophy of "panpsychic monism"—the belief that the whole of matter has consciousness and that the end result of evolution will be one universal mind linking everything—the non-Darwinian holism is manifest. Not only is the main creative work in evolution done by genetic drift, as non-Darwinian an idea as one could possibly have (as Ronald Fisher, the great English population geneticist, repeatedly pointed out), but when selection does then kick in, it is group selection not individual selection: "Selection between the genetic systems of local populations of a species . . . has been perhaps the greatest creative factor of all in making possible selection of genetic systems as wholes in place of mere selection according to the net effects of alleles" (Wright 1945, 396). And of course the whole process is one of losing and then regaining equilibrium. Obviously, Wright was taught by Henderson and fell under his spell. He wrote to his brother Quincy that he "was always very much impressed with Henderson's ideas," and acknowledged explicitly the direct influence of Spencer: "I found him a very stimulating lecturer and got lots of ideas from him, 'condition of dynamic equilibrium' etc." (Ruse 1996, quoting unpublished letters).

The Spencerian philosophy that informed Wright's thinking was a vital factor in his ready acceptance at Chicago and, clearly, given Wright's integration into the group, it was no longer possible for an ecologist to continue to ignore evolutionary factors. Reinforcing this, a new hire appeared on the scene, Alfred Emerson. Trained at Cornell as an evolutionary entomologist, specializing in the termites, his greatest intellectual debt was probably to Wheeler, with whom he interacted when working in the tropics. Defining the biological individual as "a living entity exhibiting a certain dynamic equilibrium and maintaining a relative stability in time and space," and (the language alerts us) paying homage to Spencer among others, and "without attempting to minimize the importance of the study of the parts at any holistic level," Emerson easily and enthusiastically endorsed the idea that "the animal society [is] a superorganism" (Emerson 1939, 182–83). Thus, for instance, just as we find a division of labor operating significantly at the level of the regular organism, so also in groups (naturally, Emerson's examples tended to be drawn from the social insects) we find a corresponding division of labor.

> The castes of social insects show a striking parallelism to cellular division of labor. The reproductive castes concentrate upon the function of maintaining the species and establishing new superorganismic units, . . . thus paralleling the gametes. They also produce the sterile castes, thus paralleling tissue primordia. The workers are the nutritive caste. They might be analogized with the gastrovascular system of the organism. They collect the stored energy from the environment,

comminute the particles, digest the materials, transport the substances to other castes and young, and through direct feeding supply the other units in the colony with requisite energy. . . . They also transport waste products to the exterior. (Emerson 1939, 183)

And so on and so forth. The workers are even to be compared to the outer skins or shells of organisms, given that they "perform another function through their shelter building activities. The nests establish buffer lines between the internal and external environment, thus helping to maintain favorable nest environments. In as much as shelters are usually non-living, the worker might be analogized with the shell-secreting cytoplasm of the protozoan or the shell-secreting tissue of the mollusc" (Emerson 1939, 183–84). What is important and to be expected is that Emerson wanted to locate all of this within a firm evolutionary history, and by the late 1930s, with the ongoing successes of the Darwin-Mendel synthesis, he made it clear that natural selection played a vital role. However, its role was inherently holistic, in the sense that selection could and did work on the group—the superorganism—as well as the regular individual. "Thus we find that the important ecological principle of natural selection acts upon the integrated organism, superorganism or population" (197). Immediately, Sewall Wright's authority was invoked. He was quoted as saying, "In this dependence on balance the species is like a living organism" (197, quoting Wright 1932, 365). Note again the move beyond Darwin. Social insects may have been the paradigm, but apparently the organic model can extend right up to the species.

All of this came together in a major textbook, *Principles of Animal Ecology*, that Allee and Emerson authored together with two colleagues and a former student. Starting with the environment, they moved on to populations, which they viewed as integrated entities, not just the aggregations that Allee discussed two decades earlier. They wrote, "The reality and usefulness of the population as an ecological unit were apparent to us when we outlined the present book, and our subsequent work has reinforced our conviction of the importance of the principles that center on the population. We view the population system, whether intraspecies or interspecies, as a biological entity of fundamental importance" (Allee et al. 1949, 6). They explicitly endorsed holism, and, while they noted that they were scientists and not philosophers, they made it clear that emergence was a favored notion. "In essence, emergent evolution emphasises the basic necessity for the study of wholes, as contrasted to the study of parts, and adds a certain dignity to synthetic sciences. Biology is the study of the properties of whole systems as well as of parts, and ecology, among the various subsciences of biology tends to be holistic in its approach" (693). Evolution was a major topic of discussion, and, in explanation, they introduced and endorsed group selection. "The existence of complex internal adaptation between parts of an organism or population, with division of labour and integration within the whole system, is explicable only through the action of selection upon whole units from the lowest to the highest. Mostly, these integrated levels would not exist as entities unless selection acted upon each whole system" (684). Given Allee's role as lead author, it is no surprise to find analogies suggested between lower organisms and humans. Although human social evolution is said to be "beyond the scope of this book," the authors nevertheless noted that "the social scientist may find . . . significant parallels in biological and social evolutionary mechanisms" (693). Undoubtedly aware of the dreadful political systems of Europe in the decades before they were writing (1949), the authors warned that none of their statements should be taken as justification for fascism or similar totalitarian systems.

THE ORGANICIST VIEWPOINT

Organicism passed from Schelling to Spencer, from Spencer to Harvard, from Harvard to Chicago. The claim is not that every twentieth-century scientist, specifically every life scientist, was a holist, an organicist. That is certainly not true. It was not true of Darwin, nor of those in his direct tradition down to this day.

They are mechanist-reductionists, and they are not alone. All of the groups discussed in this chapter had their critics. Many refused to buy into the Harvard philosophy. Even as they published, the work of Allee and his coauthors was often judged to be dated. Their thinking was rooted in the science and philosophy of the early decades of the twentieth century. The Second World War brought a new society and new ideas, and in ecology this was reflected in the influence of cybernetics on the approach of the English-born Yale ecologist G. Evelyn Hutchinson. Not only did he strive to bring quantitative thinking to the subject, but his thinking (and that of his students) was far more mechanistic, much influenced by the feedback ("directively organized") machines that had emerged from the Second World War (Mitman 1992).

Rather, the claim is that this organicist strand of thought, this philosophy, has always been present in the sciences, especially in the biological sciences. It may have been a minor theme, but it was never on the fringe. It was to be found in the highest places—Harvard and Chicago for a start. Moreover, it persists. Embryologists and paleontologists in particular have often been inclined this way. Embryology impresses in the way that a fertilized egg seems to contain complete instructions for producing an adult—a clear example of self-organization—and paleontologists, being one step back from causes, frequently feel little need to invoke selection and adaptation as tools of understanding. The work of Stephen Jay Gould was a paradigm. Much in the Harvard tradition, he was a leader of the campaign against orthodox Darwinism. Gould was famous (some would say notorious) for his paleontological theory (devised with fellow scientist Niles Eldredge) of punctuated equilibrium (Eldredge and Gould 1972). This theory describes the course of evolutionary history as one of inaction (stasis) interrupted by occasional sharp and rapid moves to another form. It contrasts with Darwinian "phyletic gradualism," which views the course of evolution as regular and smooth, with many intermediates between different forms. The term *punctuated equilibrium* at once suggests Spencer's "dynamic equilibrium"—and the connection is there, albeit indirectly, through Sewall Wright. The

Eldredge-Gould theory is based on an implication (the "founder principle") of the (Spencerian) shifting balance theory suggesting that rapid change occurs first in isolated populations and only later in the larger group. In support of this theory, in the name of anti-reductionism, Gould (with his Harvard colleague, geneticist Richard Lewontin) launched a major attack on Darwinian adaptation, since, for Darwin, evolution virtually had to be gradual to avoid the danger of losing adaptive focus. In their celebrated paper, "The Spandrels of San Marco and the Panglossian Paradigm: A Critique of the Adaptationist Programme" (1979), they argued against the "faith in the power of natural selection as an optimizing agent," something that "proceeds by breaking an organism into unitary 'traits' and proposing an adaptive story for each considered separately." Often, they argued, organic features are like the spandrels (triangular areas) found at the tops of columns in medieval churches. They may look as though they have an adaptive function, and perhaps (as in the church in Venice) they are now used for some purpose (in Venice, to offer mosaic portraits of the evangelists; see fig. 16), but really they are just nonadaptive by-products of the evolutionary process.

Gould and Lewontin wanted "to reassert a competing notion (long popular in continental Europe) that organisms must be analyzed as integrated wholes, with *baupläne* [groundplans or archetypes] so constrained by phyletic heritage, pathways of development, and general architecture that the constraints themselves become more interesting and more important in delimiting pathways of change than the selective force that may mediate change when it occurs" (1979, 581). Shades of Agassiz, and back to the Romantics. As we might expect, even though Gould and Lewontin criticized pan-adaptationism, both were receptive to notions of selection working at levels higher than the individual. Gould, as noted earlier, proposed a notion of "species selection" that operated at a high level and was only loosely connected to adaptation. Gould's science was clearly being driven not just by the desire to make room for his own non-Darwinian paleontological theory, but also by moral concerns. Again and again he (and Lewontin) expressed concerns that pure Darwinism might foster prejudice and

Figure 16. The Spandrels of San Marco cathedral
in Venice—a supposed example of something that
looks adaptive but actually has no structural function.
(Reprinted from S. J. Gould and R. C. Lewontin,
"The Spandrels of San Marco and the Panglossian Paradigm: A Critique of the Adaptationist Programme,"
Proceedings of the Royal Society of London, ser. B: *Biological Sciences* 205 [1979]: 581–98, by permission of
the Royal Society.)

racism by implying that humans, who have dominated others in the
struggle for existence, are more evolved (and thus of greater worth)
than others—men over women, white over black, gentile over Jew.
The reductionism seen at the heart of Darwinism—"atomization
plus optimizing selection on parts"—tends all too readily to train

a spotlight on features such as intelligence and diligence that are supposedly promoted by selection and found only in certain superior types of human beings.

The focus of the Gould/Lewontin attack was their Harvard colleague, Edward O. Wilson, the world's leading expert on the ants. In 1975 Wilson branched out and wrote a major overview of the newly developing field of the evolution of social behavior. As a climax to his book, *Sociobiology: The New Synthesis*, Wilson smoothly applied his thinking to humankind, making all sorts of claims about male dominance, female "coyness," and similar psychological and sociological facts. In their criticisms, Gould and Lewontin, taking Wilson's thinking to be deeply embedded in Darwinian theory, aimed to show that his claims about humans had to be false because his general theory was false. This was actually a wonderful case of mistaken interpretation, because in major respects Wilson shared their philosophy! His teacher was Frank Carpenter, who studied under Wheeler—so that he is the intellectual grandchild of the Harvard holists. Like them, Wilson is a first-rate scientist and shares almost all of their philosophical and social concerns, including an enthusiasm for Herbert Spencer. ("Great man, Mike!" he said to me, "Great man!")

Before he turned to social behavior, Wilson was overt about this enthusiasm; with ecologist Robert MacArthur he had devised a theory of island biogeography that made return to equilibrium absolutely central. In his subsequent overview of social behavior, the discerning reader could see the organicist philosophy at work. To the horror of the purists, Wilson described kin selection as a form of group selection, and this veering from the Darwinian norm has become increasingly clear in the more than three decades since then. In recent work on the social insects, Wilson strongly opposes the whole selfish-gene line of thought and argues that one must regard nests as individuals, "superorganisms," units of selection in their own right (Hölldobler and Wilson 2008; Wilson and Wilson 2007). Cooperation leads to "emergent properties" that favor the group. "Consider genetic variation of traits such as nest construction, nest defense, provisioning the colony for food, or raiding other colonies. All of these activities provide public goods

at private expense. All entail emergent properties based on cooperation among the colony members. Slackers are more fit than
solid citizens within a single colony, but colonies with more solid
citizens have the advantage of the group level" (Wilson and Wilson 2007, 341).

Note that despite superficial similarity—Darwin and Wilson
both regarding nests as individuals—their intent could not be more
different. Darwin took the nest, the family, as the unit of selection because, ignorant of modern genetics, he could not otherwise
account for social insects. He was still thinking at an individual
level (conceiving the nest as made of parts not members), and this
shows in his refusal to extend selection to higher units like populations and species. Also, for the same reason, he was wary about
equilibrium. Wilson deliberately turns his back on such lower levels of selection as kin selection—he does think in terms of higher
units (conceiving the nest as a group of cooperators)—and, as a
good Spencerian, he thinks equilibrium to be something evolution
cherishes in its own right.

Obviously, Herbert Spencer is not solely responsible for today's
holism. As noted, in America particularly, first the transcendentalists and later other important figures, such as Agassiz, are linked
to early German Romanticism. As an undergraduate, Peirce read
Kant's *Critique of Pure Reason* (in German) and always acknowledged his deep debt to Hegel. Most important, in the second part
of the nineteenth century German science outpaced that of other
countries, and many young Americans went to German universities to engage in the intellectual excitement. Of our Harvard scientists, Henderson had two years of advanced training in Strasbourg (part of Germany after the Franco-Prussian war); Cannon
studied at Harvard with Henry Bowditch, who was a student of
Leipzig physiologist Carl Ludwig; and Wheeler attended a German high school in Milwaukee and later studied under two scientists born and trained in Germany, Georg Baur and Anton Dohrn.
And there were more recent influences. Gould had obviously im-
mersed himself deeply in German thought and always had a soft
spot for Louis Agassiz, although he and Lewontin both held chairs
named after Agassiz's evolutionist son, Alexander. Wilson perhaps

is more diffuse—amorphous almost. Yet, again and again in his writings themes and visions emerge showing that some traditions persist. Echoing the moral heartiness of his spiritual forefathers, Ralph Waldo Emerson and his fellow Concord transcendentalists ("Slackers!" "Solid citizens!"), Wilson carries one back to a New England prep-school headmaster's Sunday evening sermon on the parable of the Prodigal Son.

SUMMING UP

Organicism is a sturdy plant, and we review here what distinguishes this view of life from that of the mechanists. The organicists see whole entities as emerging from the parts, and these are not entirely explicable simply in terms of the parts. They are integrated unities that in some way imply balance, harmony, equilibrium, and a general feeling of worth. The easiest way to distinguish the two philosophies, at least in the world of organisms, would be to say that the mechanist-reductionists are convinced that natural selection works at the level of the individual—one against all in some sense—and that everything, including harmony and altruism, can be explained in such terms. The organicists are more inclined to see selection working at the group level—populations, species, or even higher taxa—and hence the group may be regarded as a whole that is somehow greater than the sum of its parts and cannot be entirely explained in terms of its components. There is clearly some truth in this view of things, but in a way it is secondary to a deeper concern. The organicists see an integrative aspect in nature that operates outside of or beyond selection. If, like Allee, they are forced to put things in evolutionary terms, then group selection comes into play, but it is secondary to the basic way of nature. This is the real mark of organicism. There is something wholesome about nature that the hard-line Darwinian misses. Significant differences in philosophy, based largely on different traditions, led to real differences in the science.

6

HYLOZOISM

We turn directly in this chapter to those (excluding Lovelock and Margulis) who accept that the world is living. In order to locate and understand these people and their thinking it is helpful to make some distinctions and give definitions. In the last two chapters we have been dealing with thinkers who would certainly want their science to be considered as "professional." By this I mean science that satisfies the kinds of epistemic criteria that we associate with genuine and sustained efforts to find the truth about empirical reality (McMullin 1983; Ruse 1996). Such work is consistent with the facts; it binds together different aspects of the subject being studied and tries to mesh smoothly with other branches of science; it is both explanatory and yields predictions that can be tested. Cultural, social, moral values may wield some influence, but even the holists would be wary of bringing them too blatantly into the theorizing. Science is about what is rather than what we want or would like to be the case—it has moved beyond the Scientific Revolution and forsworn teleology.

What are the alternatives to professional science? One obvious candidate is what one might call "popular" science, which is aimed at the general public and appears in the science sections of major newspapers and magazines, on television, and in film. It is also to be found in the public sections of science (including natural history) museums. It is perfectly respectable—indeed, it is often produced by professional scientists—but it does not have the central aims of the professional in the sense of trying to push back the limits of knowledge. Rather, it is intended to tell people—often without the mathematics and the technical language—about what is going on in the professional world. It cannot be simply a function of cultural or other values, but we do not find the same proscription on making such values evident and providing reasons for pointing in certain directions rather than others. Professional scientists sometimes use popular science as a way to gain acceptance by their fellow professionals, hoping that popularizing their ideas will force other professionals to take heed and consider their work in the professional realm. Stephen Jay Gould was a master of this. He pushed his paleontological hypotheses in the public realm, intending to create such a stir as to force his fellow professionals take notice of what he was about (Ruse 1999b). Jim Lovelock also played this game, although the attention he received from professionals was not quite the kind he desired. At best, it is a dangerous strategy, because many professionals think that any popularization of their subject is an act of betrayal—rather like freemasons revealing the secrets of the temple.

Popular science is not the only category other than professional science. A third category is significant to our discussion but considerably more difficult to characterize. This is activity that—as we saw in the case of Rupert Sheldrake—gets labeled as "pseudo" (sometimes "crank") science. Generally, it is not produced by professional scientists (although it has been), and it is usually thought not to take sufficiently seriously the epistemic conditions that are binding on professional science. Consistency, predictive fertility, willingness to concede in the face of the evidence—these are not always cherished by pseudoscience. Conversely, far from being excluded, values are almost mandatory. The term *pseudoscience* seems to have originated in America in the 1820s (Thurs and Numbers 2011). It soon spread, and by 1843 one finds the French physiologist François Magendie using it to refer to phrenology, the supposed science that reads psychological features from the shape of the skull. But the idea predates this. In 1784, Benjamin Franklin chaired a committee for the King of France (Louis XVI) that was to evaluate mesmerism, which was based on claims about bodily magnetism, and found unambiguously that it was what we would now call pseudoscience (Ruse 1996). Both phrenology and mesmerism are deeply impregnated with value assumptions about human nature (Shapin 1975). Heated debates can arise about whether or not to label something as pseudoscience. Some would add chiropractic; others would not. This suggests that deciding if the epistemic criteria of science are really that crucial and all-determining can be an ambiguous process. As many have pointed out, professional science itself is often somewhat dicey on these scores, and (whatever others may say) pseudoscientists often claim indignantly that they do take the criteria seriously. One suspects therefore that the notion of pseudoscience is at least partly psychological or sociological: pseudoscience is whatever some people—often professional scientists or (as in the Sheldrake case) others speaking on behalf of professional science—call pseudoscience. As we shall see, this is not quite as trivial a point as we might assume (Ruse 1988; 2013a).

There is often some overlap between religion (and perhaps philosophy) and pseudoscience, and it can be difficult (and perhaps not that significant) to distinguish the two—especially when one senses more than a hint of teleology, of purpose, from one and all. In a sense, religion is not really competing with professional science, whereas pseudoscience is. Consider, for example, the story of Noah's Flood. If one believes it because one is a literalist about the Bible and is simply indifferent to the geological evidence (God could have cleaned things up after the event), then one is not really in the realm of science at all. This is a religious claim that others might reject and that scientists might reject with scorn. But it does not really compete with professional science. However, if one insists with the creation scientists that the evidence does point to the Flood (the fossil record is explained not through evolution but as an artifact of the order in which organisms were drowned), then one is in the realm of pseudoscience. Consistency with other sciences, among other things, is being ignored. But given that (as here) the motivation of the pseudoscience is really cultural, a defense of a particular form of religion, and given that (as here) pseudoscience and religion can and do often overlap, the opposition is often against the very system of claims and expends little effort on precise labeling.

All of these alternative categories—popular science, pseudoscience, religion (and philosophy)—are discussed in this chapter, for we are dealing with people who in some respects stand outside the mainstream of professional science and are, in varying degrees, indifferent to this fact. Many of them stand outside the mainstream of religion and are also, in varying degrees, indifferent to this fact as well. It is impossible to cover everyone and everything but, as before, I hope to give a true view of the nature and vigor of this mode of thinking. The people discussed are not chosen randomly but with an eye to our overall discussion. The ideas and people may sometimes seem rather strange—even very strange indeed—but do not underestimate either their sincerity or their appeal. Although strange, the ideas are not entirely unfamiliar. They are not only children of their time but are also very much part of the Platonic tradition we have been following. This means that one significant characteristic of the people to be discussed is the extent to which they intermingle fact and value, and often think happily in terms of overall ends or purposes; they would consider not doing so to be not only impossible but morally wrong-headed.

ANTHROPOSOPHY

Born in the Austrian-Hungarian empire (in what is today part of Croatia), the mystical polymath Rudolf Steiner (1861–1925) emerged from the fermenting broth of German culture, particularly German Romanticism (Easton 1980; Lachman 2007). As a young man, Steiner labored in the Goethe archives in Jena, working on the definitive edition of the great man's writings. Steiner's assignment covered the science, and this left a deep and lasting impression. He always embraced the holistic morphology that downplayed adaptation and played up links, parallels, reflections of the whole. Steiner was also much taken with the philosophical musings of the well-known, *Naturphilosoph*-influenced German evolutionist Ernst Haeckel. In the spirit of the Romantics, Haeckel declared, "I am entirely at one . . . as to that unifying conception of nature as a whole which we designate in a single word as Monism. By this we unambiguously express our conviction that there lives 'one spirit in all things,' and that the whole cognizable world is constituted, and has been developed, in accordance with one common fundamental law. We emphasise by it, in particular, the essential unity of inorganic and organic nature, the latter having been evolved from the former only at a relatively late period" (Haeckel 1895, 3). Steiner enthusiastically responded, "The essence of *Monism* consists in the assumption that all occurrences in the world, from the simplest mechanical ones upwards to the highest human intellectual creations, evolve themselves naturally in the same sense, and that everything which is called in for the explanation of appearances, must be sought *within* that same world" (Steiner 1914a, 110–11).

This all fit nicely with a move Steiner was making from pure science to philosophy. He wrote a doctoral thesis on the arch-idealist Fichte, another thinker who bolstered Steiner's belief that in some sense everything is throbbing with life. Schelling also was a significant influence, and late in life Steiner spoke of the philosopher's "important inspirations" in a lecture given at Dornach, Switzerland, September 16, 1924 (Steiner 1957, Lecture 6). In the same lecture he said that as a young man he "again and again returned to Schelling," and Steiner took great care to stress the neo-Platonic aspect of Schelling's thinking, which he thought "deeply penetrating." Then came what was surely the overriding influence on Steiner's subsequent career. To set the context, bear in mind that by the end of the nineteenth century, although Germany (where Steiner was now living) was finally unified and changing rapidly from an essentially rural nation to one that was powerful and industrial, many were uncomfortable and resisting modernity. Their reaction was evident, for example, in the obsession with things of a volkish or quasi-medieval nature, such as fairy stories and the operas of Wagner. There was also an interest in the occult and fantastic, stories of magical powers and of ways of seeing into the mysteries of life. Dropped into this mélange, adapting and growing rapidly, was an American import: theosophy. Theosophy was the brainchild of one of the most interesting adventurers (some would say charlatans) of the nineteenth century, Madame Helena Blavatsky, a Russian-born, American-naturalized psychic and spiritualist, who spent much of her later life in India and England. Theosophy is a strange, esoteric combination of religion (especially Eastern religion) and neo-Platonism, particularly a kind that sprang up in the sixteenth century that involved mysteries known only to initiates and often a belief in souls or spirits that permeated the universe, and fostered warm feelings toward the whole of humankind (Cranston 1994).

Steiner, who by the beginning of the twentieth century was openly claiming to have clairvoyant powers of his own, was one of many who embraced theosophy, with so much enthusiasm that he became head of the German branch of the movement. This lasted for more than a decade until Steiner broke with the theosophists and articulated his own world picture, which became the basis for the movement he founded: "anthroposophy." The details are extraordinarily complex and expounded at mind-numbing, Germanic length in the many lectures (more than six thousand) given and recorded (in more than three hundred volumes) by devoted followers, but the basic principles are relatively simple (Steiner 1914b) and, although strange, not entirely unfamiliar. Start with the individual human. Echoing Hindu thinking, Steiner says we are a bit like an artichoke. At the center is the essential soul, the ego. Around this center are layers. The outermost layer is the physical body—it is this that we have at the beginning of life and that we have until we die, and it is this that rots after death. Next there is the etheric body. Plants have this and only this (inside the physical body), whereas animals and humans have more. The etheric body is responsible for keeping us going, as it were. This is the physiological part of the human being. Within the etheric body is the astral body. This enables the movement and consciousness that we humans share with animals. Finally, unique to humans, we have the subtle body that yields self-consciousness—the ego at the center. Fusing this vision with evolution, we have at birth just the physical body. Then the etheric body develops alone until we are about seven, the age of the coming of the second teeth. After that, the astral body develops until adolescence, and finally, through the later teens, the subtle body develops until we are about twenty-one and finished.

This psychology of development is the foundation of Rudolf Steiner's great popularity and influence today. Just after the First World War, Steiner was asked by a wealthy follower and industrialist—the owner of the Waldorf tobacco factory—to found a school for the children of his workers. This Steiner did, drawing directly on his own theory. Until the age of seven, the child was to be treated, if not exactly as a vegetable, certainly not as a person with independent powers of thought. Reading was not encouraged, but the teacher told fairy stories and the like. (The unkind critic might suggest that this had at least as much to do with German authoritarianism as with anthroposophy.) After the age of seven, the child's education reflected the developing consciousness up to the age of adolescence. After that, more philosophical ideas could be introduced. "Waldorf" education today is growing at an explosive rate; the number of schools, from America to (of all places) Israel, has increased fivefold in the past three decades.

THE COSMIC WORLD

As we dig further into Steiner's system, we find that it diverges more radically from conventional thinking. Grabbing ideas from past and present, near and far, Steiner offered up a remarkable gallimaufry of sweeping and fantastical claims, from positing an age when humans all lived in Atlantis, where now we find the Atlantic—an idea, incidentally, to be found in the *Timaeus*—to his view of the coming of Christ. He affirms that there were actually two Jesus children who fused together to make the Christ, a bit like the Skeksis and Mystics who come together to make the urSkeks in the movie *The Dark Crystal*. Angels also get involved, taking human souls from one planetary sojourn to another—one lasting relic from theosophy was a belief in reincarnation—improving us until we are ready for Earth. "Just as humankind is being brought to a higher level by the Earth phase of evolution, this was also the case in each of the earlier planetary incarnations, since a human element was already present even during the first of these incarnations. This is why we can shed light on the essential nature of present-day human beings by tracing our evolution back to the very distant past, to the first planetary incarnation" (Steiner 1914b, 128).

Presupposed in Steiner's philosophy, no doubt due to the influence of Haeckel's monism, is a central and significant Platonic or neo-Platonic vision of Earth (or any planet for that matter) as a living entity. This is the basis for a remarkable foray into the science of agriculture. In the early years of the Weimar Republic, Steiner was invited to speak on the subject by a group of devotees, and he laid down the basic ideas for the "biodynamic approach" to farming (Steiner 1924), which (as with his views on education) seems today to be attracting adherents by leaps and bounds. In some respects, it is a variant of straightforward organic gardening—manures from farm animals, natural approaches to pest control, concern about taste over ease of manipulation, and so forth. Surface similarity however should not conceal the fact that its prescriptions are deeply embedded in anthroposophical theory. A key premise is the need to spray fields with special "preparations" at times determined by the conjunctions of the moon and planets, thus assuring the friendly involvement of spirits from elsewhere in the universe. One preparation ("500") involves stuffing a cow's horn with manure, burying it for the winter, and then mixing it with water stirred in a special fashion. Why a cow horn? (Steiner rejects the use of inferior substitutes like stags' horns.) The horn is the instrument through which the cow takes in astral forces that facilitate its digestion. "Thus in the horn you have something well adapted by its inherent nature, to ray back the living and astral properties into the inner life. In the horn you have something radiating life—nay, even radiating astrality. It is so indeed: if you could crawl about inside the living body of a cow—if you were there inside the belly of the cow—you would *smell* how the astral life and a living vitality pours inward from the horns" (73).

In short, proper agricultural practice is more than a matter of getting the right molecules in the right places. It involves awareness of astral or occult forces, emanating from the other planets and affecting what goes on here on Earth. "The activities above the Earth are immediately dependent on [the] Moon, Mercury and Venus supplementing and modifying the influences of the Sun. The so-called "planets near the Earth" extend their influences to all that is *above* the Earth's surface. On the other hand, the distant planets—those that revolve outside the circuit of the Sun [*sic*] work upon all that is *beneath* the Earth's surface, assisting those influences which the Sun exercises from below the Earth." The practical implications follow at once: "Thus, so far as plant-growth is concerned, we must look for the influences of the distant heavens *beneath*, and of the Earth's immediate cosmic environment *above* the Earth's surface" (Steiner 1924, 30–31).

A biodynamic agricultural enthusiast is therefore quite unmoved by a cynic who points out that the practice also embraces the ideas of homeopathy and that (given the minute levels of the doses) this is totally ineffective. The follower of Steiner brushes off such objections. We are simply not thinking at the conventional material level at all. Rather, anthroposophical agriculture works because everything, at every level, is throbbing with life. For Steiner, we have the individual organisms in the farm. Then the farm itself, if run properly, is an organism. "A farm is true to its essential nature, in the best sense of the word, if it is conceived as a kind of individual entity in itself—a self-contained individuality. Every farm should approximate to this condition" (Steiner 1924, 29). And finally this is true of Earth itself. Although it turns out that everything is inside out, literally! "Taking our start from a study of the earthly soil, we must indeed observe that the surface of the Earth is a kind of organ in that organism which reveals itself through the growth of Nature. The Earth's surface is a real organism, which—if you will—you may compare to the human diaphragm." Important are the head and the stomach. "If from this point of view we now compare the Earth's surface with the human diaphragm, then we must say: In the individuality with which we are here concerned, the head is *beneath* the surface of the Earth, while we, with all the animals, are living in the creature's belly! Whatever is *above* the Earth, belongs in truth to the intestines of the 'agricultural individuality,' if we may coin the phrase" (30).

ECOLOGICAL PHILOSOPHIES

Anthroposophy is obviously well beyond the bounds of professional science. Moreover, although, given the success of Waldorf education, anthroposophy is very popular in some respects, we are not really dealing with popular science in the sense being used here. We have rather a mishmash of religion on the one hand and pseudoscience on the other, as critics have pointed out (e.g., Shermer 2002, 32). It is hard to tell where one ends and the other begins, but for our purposes it is not really important. This is not to say that employing the idea of the world as an organism automatically takes one into the realm of religion or philosophy or (more dangerously) into the realm of pseudoscience. As presented here, however, it certainly becomes tainted with the odor of pseudo- or crank science. Yet nothing precludes anthroposophists (or adherents of any other pseudoscience) from reaching out and relating to others—possibly even to professional scientists and certainly to those in the domain of popular science. Shared values can make for strange bedfellows. Specifically pertinent here is that Steiner's followers take a deep interest in environmental matters, feeling that they have unique insights to bring to the problems. But even they realize that they are part of a larger, broader movement, and it is no surprise that American anthroposophists revere the memory of the American transcendentalists, more generally for their debt to German Romanticism and more specifically for the movement's love of and concern with nature.

Steiner enthusiasts are not alone in this respect for transcendentalism, for it is surely one of the most significant inspirations for what today is known (somewhat to the irritation of professionals) as *ecology*, meaning now not the formal science discussed in the last chapter but something more in the realm of popular science (in our terms), an idea very much in the public domain and rooted in a more general philosophical concern with the environment and related social issues. The two senses of *ecology* are not entirely separate, for the more philosophical-social sense certainly draws on professional science, and, given the holism of professional science, it would be odd if organicism (perhaps of a stronger kind) was absent from the more philosophical sense of ecology.

Let us start gently in territory that is familiar to anyone who has been through the American school system. Much more than Emerson, his younger contemporary Henry David Thoreau loved nature for its own sake, and spent much time outdoors enjoying, recording, and responding to it. His life and thought testified to the Romantic sense that everything is alive. Rejoicing over the coming of spring and how new life springs up from within, Thoreau wrote,

> There is nothing inorganic. These foliaceous heaps lie along the bank like the slag of a furnace, showing that Nature is "in full blast" within. The earth is not a mere fragment of dead history, stratum upon stratum like the leaves of a book, to be studied by geologists and antiquaries chiefly, but living poetry like the leaves of a tree, which precede flowers and fruit—not a fossil earth, but a living earth; compared with whose great central life all animal and vegetable life is merely parasitic. Its throes will heave our exuviae from their graves. (Thoreau 1854, sec. 2, 476)

Partly because of Thoreau's influence, given the love of his readers for his spirit and his joy in the world, organic metaphors about Earth are part of the American heritage. John Muir, who founded the Sierra Club and was the most important figure promoting nature in America in the second half of the nineteenth century, was warmly embraced by the transcendentalists (Worster 2008). Ralph Waldo Emerson included Muir in his list of about twenty people, designated "my men," and there are good grounds for this. Muir's philosophy-cum-theology was a kind of pantheism: "Beauty is God, and what shall we say of God that we may not say of Beauty?" (Worster 2008, 208). Moreover, although Muir became a Darwinian, he preferred the softer, cooperative version. "I never saw one drop of blood, one red stain on all this wilderness. Even death is in harmony here" (Muir 1966, 93). Organicism, even full-blown hylozoism, is evident here, although for Muir it was more the result of shared sources than direct import from the transcendentalists. He was born in Scotland. His primary influences came from there, and above all he was inspired by the poet William Wordsworth: "a spirit, that impels / All thinking things, all objects of all thought, / And rolls through all things" (*Tintern Abbey*, 1789).

Moving to the twentieth century, we find the same themes. The naturalist Aldo Leopold, a long-time promoter of game and land management from the University of Wisconsin, is probably the only person who has become as beloved in memory as Thoreau. His *A Sand County Almanac*, published posthumously (in 1949) after he died working a managed forest fire, has made him a hero of the environmentalist movement of the past half century (Meine 1988; Newton 2006). The notion of Earth as an organism lies at the heart of his thinking. In 1923, in a (then-unpublished) essay, Leopold was arguing that many of us "have felt intuitively that there existed between man and the earth a closer and deeper relation than would necessarily follow the mechanistic conception of the earth as our physical provider and abiding place" (Leopold 1979, 139). He went on to say, "Philosophy, then, suggests one reason why we can not destroy the earth with moral impunity; namely, that the "dead" earth is an organism possessing a certain kind and degree of life, which we intuitively respect as such" (140), a position he held until the end. "All ethics so far evolved rest upon a single premise: that the individual is a member of a community of interdependent parts. His instincts prompt him to compete for his place in that community, but his ethics prompt him also to co-operate (perhaps in order that there may be a place to compete for)." He continued, "The land ethic simply enlarges the boundaries of the community to include soils, waters, plants, and animals, or collectively: the land." (Leopold 1949, 204).

The transcendentalists and Muir figure in Leopold's background—he even included Emerson in copied-out quotations from great men that he made as a schoolboy—but the direct influence seems to have been P. D. Ouspenski, the Russian-born sometime disciple of the esoteric philosopher G. I. Gurdjieff. And when we examine Ouspensky's so-called masterwork, *Tertium Organum: A Key to the Enigmas of the World* (1912), we find that his inspiration for thoughts about world souls comes (apparently second-hand, through an exposition by William James) from the nineteenth-century German experimental psychologist Gustav Fechner. "The earth-soul he passionately believes in; he treats the earth as our special human guardian angel; we can pray to the earth as men pray to their saints" (Ouspensky 1912, 189, quoting William James writing about Fechner). Ouspensky commented as follows: "Logically we must either recognize life and rationality in everything, in all 'dead nature,' or deny them completely, even IN OURSELVES" (191). Fechner, writing in the later part of the century (that is, at the same time as Haeckel), was an ardent monist, and a major philosophical influence on his life, as we might expect, was his fellow countryman, Friedrich Schelling. Whether or not Leopold had even heard of the German idealist, it is obviously hard to keep a good idea down.

RACHEL CARSON AND *SILENT SPRING*

Aldo Leopold is the one who is loved, but it was the marine biologist and science writer Rachel Carson who made environmentalism into an idea whose time had come (Lear 1997). Already a well-known figure because of her powerful books about the ocean and its shores—*Under the Sea-Wind* (1941), *The Sea Around Us* (1951), and *The Edge of the Sea* (1955)—her devastating *Silent Spring* (1962), in which she recorded the horrendous effects of the insecticide DDT on America's wildlife, brought home to everyone how fragile our planet is and how easy it is for us to destroy it. Chapter after chapter pounds away at her theme. We learn about insecticides, particularly about chlorinated hydrocarbons, which include DDT (dichloro-diphenyl-trichloro-ethane). We learn how heavily they are being used and how they are spread through the earth, particularly through its waters. We learn how disruptive this is for the world's wildlife and how difficult it is to object to or counter any of it. "Under the philosophy that now seems to guide our destinies, nothing must get in the way of the man with the spray gun. The incidental victims of this crusade against insects count as nothing; if robins, racoons, cats or even livestock happen to inhabit the same bit of earth as the target insects and to be hit by the rain of insect-killing poisons no one must protest" (Carson 1962, 85–86). Nor are we humans safe. In our houses, we are allowed to use pesticides that can send people into convulsions. "As things stand now, we are in little better position than the guests of the Borgias" (184). Cancer is almost inevitable, and we continue to endanger ourselves so long as curing is promoted over elimination. "The most determined effort should be made to eliminate those carcinogens that now contaminate our food, our water supplies, and our atmosphere, because they provide the most dangerous type of contact—minute exposures, repeated over and over throughout the years" (242). In the end, it all breaks down. Insects develop natural resistances to insecticides. More and more alien substances are used, and they become less and less effective. We destroy ourselves and our world, and in the end we lose. Yet Carson is not entirely negative. On the one hand, rather than calling for an outright ban on insecticides, she encourages more moderate and reasoned use. On the other hand, she does call for other, more organic approaches that use natural means—the pests' own natural enemies, primarily—to control the pests that trouble us.

Rachel Carson was trained as a scientist, but she wrote for the public. In our sense of the term, she was the greatest popular scientist of the twentieth century. She wrote in striking prose, almost in poetry; she talked of serious things (and took great care to get them right) but never in a technical way; and her moral concerns were there for all to see. She was also a very skilled and politically savvy writer. Much of her writing in *Silent Spring* suggests strongly that it would be easy for Carson to base her discussion on the concept of Earth as organism—most obviously, her view of the world (its soil, its waters, its weather) and its inhabitants (plants, animals, humans) as interrelated, as one. A disturbance of one part of the world resonates throughout the rest of the world. We find language that supports these ideas: one chapter heading is "The Earth's Green Mantel"; another is "Nature Fights Back." Her values also support them: she writes of "the obligation to endure" (Carson 1962, 13). Of squirrels dying from exposure to our chemicals, she writes, "By acquiescing in an act that can cause such suffering to a living creature, who among us is not diminished as a human being?" (100) And of course there is the whole analogy between wrecking our home and wrecking our own bodies. Yet Rachel Carson simply does not go that way, even in metaphorical terms. She stays strictly away from any explicit suggestion that the world should be taken as a living being. She does not quote others on this topic, nor does she suggest it herself. Why? She knew that she was going to be attacked ferociously by the vested interests— the chemical manufacturers, the agriculturalists, the food industry generally—all of whom would find their businesses in jeopardy if her message was accepted. She knew that setting her story explicitly against the background of hylozoism would bring trouble. Her critics would seize on it and call her a philosopher or some such thing. Determined to bring her down, they would label her a crank, a purveyor of pseudoscience. She could not afford to provide any reason for this to happen.

Yet readers had little doubt that Carson was committed, if not to authentic hylozoism, at least to a deeply organicist view of nature that sees life and its abode as one integrated whole. We would expect this of one whose biological training took place around 1930 (she did some graduate work at Johns Hopkins University). She identified herself as a child of Thoreau (CC) and loved his writing about the Maine Woods: "I looked with awe at the ground I trod on, to see what the Powers had made there, the form and fashion and material of their work. This was that Earth of which we have heard, made out of Chaos and Old Night. Here was no man's garden, but the unhandselled globe. . . . There was there felt the presence of a force not bound to be kind to man. It was a place for heathenism and superstitious rites,—to be inhabited by men nearer of kin to the rocks and to wild animals than we" (Thoreau 1864, 2). You find this from the earliest of her public (magazine) writings to her last written words. All is connected, part of "that essential unity that binds life to the earth" (Carson 1955, 250). Continuing this passage that Carson wanted to have read at her funeral service, she writes, "We come to perceive life as a force as tangible as any of the physical realities of the sea, a force strong and purposeful, as incapable of being crushed or diverted from its ends as the rising tide." She goes on to say that, as we look at life, "we have an uneasy sense of the communication of some universal truth that lies just beyond our grasp." We have the quest but not the answer. "The meaning haunts and ever eludes us, and in its very pursuit we approach the ultimate mystery of Life itself" (250).

The vision was there—a vision that in Carson's final great work was even more thoroughly permeated by the themes of this chapter than many recognize, even today. In writing *Silent Spring*, Rachel Carson drew heavily on the thinking and activities of people deeply committed to Rudolf Steiner's views about nature and agriculture. Carson apparently never embraced the wilder aspects of Steiner's metaphysics, but the concerns of the anthroposophists about the state of nature were vital. In the years leading up to the appearance of *Silent Spring*, one of Carson's most constant and informative sources was Marjorie Spock, a committed anthroposophist and ardent biodynamic gardener. Spock, "a woman of enormous courage, integrity, and indefatigable spirit who soon became one of Carson's inner circle of friends and the central point of her original research network," fed Carson all sorts of material, including material about the deadly effects of DDT (Lear 1997, 318). Evidence of this appears frequently in their voluminous correspondence: "You are my chief clipping service" (RC to MS, September 26, 1958); "The mass of your material I have here . . ."

(RC to MS, December 8, 1958); "wealth of material" (RC to MS, January 18, 1960); "excellent clippings" (RC to MS, October 11, 1961); and so on (YU).

Most significantly, in January 1958, the anthroposophical magazine *Bio-Dynamics* published an article by the leading biodynamic agriculturalist in the United States, Ehrenfried E. Pfeiffer, with the title "Do We Know What We Are Doing? DDT Spray Programs—Their Values and Dangers." Within a month, courtesy of Marjorie Spock, it was in Carson's hands, who replied with thanks, noting that it was a "gold mine" (RC to MS, March 26, 1958 [YU]). It certainly was. It described how quickly organisms build up resistance to DDT; how much of it winds up in our foodstuffs; and how it persists and builds up in the fatty tissues of animals, including ourselves. It explains that there are more biological, natural ways to combat pests, but that the chemical industry and others with vested interests oppose or ignore them. If anything, it alters or destroys the natural balance. "Studies with about 4 lbs DDT/acre as 2% water emulsion spray revealed that a single application may do more harm than good by reducing natural predators to mosquito larvae" (Pfeiffer 1958, 13). Fruits and vegetables are destroyed or their flavors ruined. The well-known effects of DDT on birds' eggs is documented. "Feeding of diets containing 0.02% DDT to breeding quail did not influence the adult birds but reduced the hatchability of eggs and the viability of chicks. Chicks showed high mortality, even when fed a DDT-free diet" (17). And we should certainly be skeptical of claims that DDT is harmless to humans. "Do we really understand the laws of balances in nature, of long range observations?" (32)

There was no attempt here to conceal the underlying philosophy. We learn that a "spiritual orientation, a philosophy, is necessary, a deeper insight, the lack of which we discover in many scientists" (Pfeiffer 1958, 8). "'Everything tries to maintain itself against damage and oppression.' This sentence contains the basic philosophy of life, of biology. 'Against damage' means that a living concept would try to stimulate the natural defenses, to enforce such growth conditions which enable the growing organism to remain protected. 'Against oppression' means that nature will always answer with ways and means to fight off brute force, the toxic impact of influence by man for instance, by building up immunities which render the continuation of toxic sprays useless" (8–9). In short, "we human beings can strive—at our best—to comprehend only, rather than to imitate or improve on higher wisdom which, after all is the foundation to which we owe our existence" (33).

Spock's connection to Steiner's philosophy was no secret. Without her consent, her farm had been sprayed by government authorities, and she took the case to court, arguing all the way up to the Supreme Court. She lost, and the judges along the way made it clear that they did not trust her experts precisely because they were into "organics"; instead the judges were convinced by traditional experts who testified that spraying carried no dangers. So, although Rachel Carson was eager to use the material furnished by the anthroposophists, she studied the court case with great care. She certainly could not afford to identify publicly with them or to give any indication that she shared their overall philosophy. She was very sensitive to potential contamination by association. In 1960, a newspaperman, William Longgood, published a book on food additives, *The Poisons in Your Food*, which received extremely critical reviews. Although Marjorie Spock pushed this book on Carson, the writer deliberately avoided reading it, noting that the very fact that Longgood had reported on the Spock lawsuit "would automatically make him a target of the New York State Department of Agriculture" (see MS to RC, February 26, 1960; RC to MS, March 14, 1960; RC to MS, May 18, 1960 [CC]). Any public acknowledgement of her friend would have opened the way for critics to label Carson a crank, a pseudoscientist. And so she remained silent. *Silent Spring* did not include one printed word of thanks or acknowledgement to Marjorie Spock or to Ehrenfried Pfeiffer (although some of his references reappear in her book).

From our perspective, we can see how Steiner's worldview, implying as it did the idea of the world as an organism, was in important respects a major influence on *Silent Spring*. We can also see even more reason why, whatever Rachel Carson did or did not say overtly in her book, others might pick up or find confirmation of such a vision and incorporate it into their thinking. And this holds for those who might themselves feel no specific allegiance to Steiner, perhaps not even knowing of him and his philosophy.

THE VISION CONTINUES

At once these predictions came true. *Silent Spring* appeared at the beginning of the sixties. Even if you ignore the iceberg beneath the water, there was quite enough above the surface to inspire and motivate the growing numbers who felt that things were going radically amiss, especially as the war in Vietnam began to heat up, confirming for many President Eisenhower's warning about the threat of the military-industrial complex. *Silent Spring*'s stand against the chemical manufacturers and fellow travelers was taken as having deep spiritual significance—an emotion that was intensified by the wholesale chemical defoliation of large parts of Vietnam in an attempt to deprive the enemy of cover.

Among those incredibly impressed and influenced by *Silent Spring* was the Norwegian philosopher Arne Næss. Carson's work spurred him to turn from topics in conventional philosophy to environmental issues, formulating a position that became known as "deep ecology," which gave priority to a concern for the world around us. In a way, it is not just the most important concern, it is the only concern. It was expressed in a multipoint manifesto whose first principle was the key: "The well being and flourishing of human and non-human life on Earth have value in themselves (synonyms: intrinsic value, inherent worth). These values are independent of the usefulness of the non-human world for human purposes" (Næss 1995b, 68). Næss stressed that the term *life* means more than just organisms, the biosphere; it means the whole ecosphere. "The term 'life' is used here in a more comprehensive non-technical way also to refer to what biologists classify as 'non-living': rivers (watersheds), landscapes, ecosystems. For supporters of deep ecology, slogans such as 'Let the river live' illustrate this broader usage so common in many cultures" (68).

This makes it seem as if the idea of world as organism is reappearing. It will furnish the ultimate basis for the intrinsic values to which appeal was being made and form the foundation of the claim that we ought to cherish the ecosphere for its own sake. In the opinion of Gaia enthusiast Fritjof Capra, mechanism—"the view of the universe as a mechanical system composed of elementary building blocks, the view of the human body as a machine, the view of life in society as a competitive struggle for existence, the belief in unlimited material progress to be achieved through economic and technological growth" (Capra 1987, 19–20)—is the enemy. It has been tried and found lacking. We are faced with a need for "radical revision" (20). But soon the deep ecology program sounds as if the revision is leading us back into familiar territory as much as forward into the unknown. Spinoza is praised, especially inasmuch as the Dutch philosopher emphasized the totality of things and the spirit-infused nature of the material world. The value-impregnated nature of existence is stressed: "We tend to say 'the world of facts,' but the separation of value from facts is, itself, mainly due to an overestimation of certain scientific traditions stemming from Galileo that confuses the *instrumental* excellence of the mechanistic world-view with its properties as a whole philosophy" (Næss 1995a, 253). Ultimately we learn that deep ecology is based on the idea that the world is an organism and, as such, worthy of care and attention. It has its own intrinsic value. "I suspect that our thinking need not proceed from the notion of living beings to that of the world, but we will conceive reality, or the world we live in, as alive in a wide, not easily defined, sense. There will then be no non-living beings to care for" (Næss 1995c, 234).

Similar sorts of thinking—including hints of purpose—characterized the "ecofeminist" movement. Again, Rachel Carson is an inspiration and role model, not least for her willingness to battle the (male-dominated) scientific-industrial power structure. "Her highly publicized conflict with the male establishment demonstrates, for the most recent generation of feminists, how patriarchy has tried to use the gendered connection of women with nature to contain their voices of opposition" (Norwood 1993, 280). Much is made of the network of women, starting with her mother, who were so important in Carson's life (although, as noted, Marjorie Spock is written out of the story). This praise for women and yet exclusion of the problematic (no matter what its actual importance) is part of the ideological myth-making of ecofeminism, namely, the essential significance—the greater significance—of the female compared to the male. This reflects the major difference between ecofeminism and deep ecology: the former puts the blame on males rather than on all of humankind. "For women making the connections between the masculinist ravaging of nature and the rape of women, Carson was a forerunner. She saw the problem for nature: the arrogance of men who conceive of nature for their own use and convenience. We see it for women: a similar arrogance which assumes that women exist for the use and convenience of men" (Hynes 1989, 55).

Yet both movements are spurred by shared concerns. Stimulated and motivated by *Silent Spring*, both engage in the war against chemicals (Merchant 1995, 150), and, clothed in appropriate language, the underlying metaphysics is the same. Ecofeminist thinking is rooted in a worldview for which a living Earth is the essential premise and starting point. "The physical rape of women by men in this culture is easily paralleled by our rapacious attitudes toward the Earth itself. She, too, is female. With no sense of consequence in the scant knowledge of harmony, we gluttonously consume and misdirect scarce planetary resources" (Razak 1990, 165). The world is in an ecological mess. Men have made it so, and part of making it so has entailed the oppression of women. Now women have the chance to take control; only then will we see improvement. The neo-Platonic resonances behind this are spelled out: "All is One, all forms of existence are comprised of one continuous dance of matter/energy arising and falling away, arising and falling away. Only the illusions of separation divide us. The experience of union with the One has been called cosmic consciousness, God consciousness, knowing the One Mind, etc." (Spretnak 1989, 127). The implications are obvious. "The planet, our mother, Grandmother Earth, is physical and therefore a spiritual, mental, and emotional being. Planets are alive, as are all their by-products or expressions, such as animals, vegetables, minerals, climatic and meteorological phenomena" (Allen 1990, 52). Believing that the Earth is just inert matter is destructive and physically disease-making to the individual (Starhawk 1990, 74).

Hear the earth sing
    of her own loveliness
    her hillock lands, her valleys
    her furrows well-watered
    her untamed wild places
She arises in you
    as you in her
Your voice becomes her voice
Sing!
(Starhawk 1990, 86)

PAGANISM

With the New Age culture and its fellow travelers, we find ourselves in murky waters where philosophy melds into pseudoscience and, as always in America, on to religion. Ecofeminists are somewhat selective about pseudoscience, having no objection to it when it meets their needs. They are much into myths about lost Edens when women led the way. This segues comfortably into quasi-religious group experiences often involving chanting (as above) and related paraphernalia (candles, robes, and so forth), not to mention other practices, including the exclusion of non-initiates (men). Even if *Silent Spring* was not overtly religious, it was the way in which the book spoke to spiritual needs that made it so very powerful and influential. This highlighted the paradox (noted in the introduction) that traditional religion was failing to satisfy and speak to the needs of the day. Indeed, religion was regarded as the problem as well as the solution. An influential article by historian Lynn White (1967) firmly laid many of the problems of the environmental world on Christian theology, especially inasmuch as the early chapters of Genesis have been taken as license from God for humans to exploit the world, both the physical and the living.

Deep ecology and ecofeminism represent attempts to put things on a better foundation, a base for those who want to take Earth and its nature and needs as central. Value lies in the Earth itself, rather than in something conferred by a Creator God. Following in the steps of theosophy, many environmentalists searching for a new faith—especially the deep ecologists—urged people to travel East. The religions of Asia were regarded as more welcoming of beliefs about the living Earth. And there may indeed be truth in this, although often people who came seeking were determined to find and (hardly any surprise) happily found what they sought. Others suggested that it was not necessary to stray so far from home to find elements of organic thinking. The native religions of North America, which take very seriously the world in which we live and our duties and obligations to it, were promising. In the words of one such believer,

> Our entire existence is of reverence. Our rituals renew the sacred harmony within us. Our every act—eating, sleeping, breathing, making love—is a ceremony reaffirming our dependence on Mother Earth and kinship with her every child. We Native Americans recognize the "spiritual" and the "physical" as one—without Westerners' dichotomies between God and humankind, God and nature, nature and humankind, we are close and intimate and warm with Mother Earth and the Great Spirit. Unlike Christian belief, which claims that our species is both inherently evil and the divinely ordained ruler of Earth, we know that, being of our sacred Mother Earth, we are sacred. (Adler 2006, 383)

The key thing about native religions is that they tend to start with Earth and move out from there to God or gods, whereas Christianity goes the other way around. Most promising of all, therefore, for many seekers was a religion that made Earth absolutely central to their theological visions—Paganism (Hutton

2001). Not only do many Pagans identify with native religions—
one adherent says of the just-quoted passage that this "statement
is close to words I heard over and over again from Neo-Pagans"
(Adler 2006, 383)—but they also connect with others discussed in
this chapter. For instance, many of the more prominent ecofeminists think of themselves as Pagans in some sense of the term.
"The core thealogy [*sic*] of Goddess religion centers around the
cycle of birth, growth, death, decay, and regeneration revealed in
every aspect of a dynamic, conscious universe. The Goddess is the
living body of a living cosmos, the awareness that infuses matter
and the energy that produces change. She is life eternally attempting to maintain itself, reproduce itself, diversify, evolve, and breed
more life; a force far more implacable than death, although death
itself is an aspect of life" (Starhawk 1979, 244).

Like the anthroposophists, Pagans may not be numerous—optimistic estimates put the upper limit today at about a million, compared to the Mormons, for instance, who (in the United States)
number more than five million—but (like anthroposophists) their
numbers are growing rapidly. Moreover, they are skilled at making public their ideas, and they have influence (especially among
the young) that is disproportionate to their numbers. Check the
philosophy section in any chain bookstore.

One could debate endlessly the extent to which the Pagans of
today have genuine continuity with the pagans of antiquity, but in
the central focus on a living Earth with its own intrinsic value there
is far more similarity than difference. To conclude this chapter,
showing just how deeply the world-as-organism idea has become
embedded in some aspects of Western (especially American) culture in the 1960s and 1970s, let me introduce one of the most
fascinating of our cast of characters—an American, born in 1942,
who was named Timothy Zell at birth but has recently taken the
name of Oberon Zell-Ravenheart (Adler 2006). This remarkable
man describes himself on his website as a "transpersonal psychologist, metaphysician, naturalist, theologian, shaman, author, artist,
sculptor, lecturer, teacher" and is also "an initiate in the Egyptian
Church of the Eternal Source" as well as "a Priest in the Fellow-
ship of Isis." This all came somewhat later, for at the beginning of

the 1960s, Tim Zell (to use his given name) was a rather conventional undergraduate at Westminster College, a liberal arts college in Missouri. Yet he and his fellows sensed that the old was passing and the new was coming. Fostered by the attack on mechanical science represented by such books as *Silent Spring*, the new dimensions for personal relations opened by such (chemical) discoveries as the pill, the readily available opportunities for various drug-induced, mind-altering or expanding experiences, and much more, the Age of Aquarius was nigh. As Bob Dylan told us all: "The times they are a-changin."

For Zell, the turning point—he really did have a road-to-Damascus experience—was reading a science-fiction novel, *Stranger in a Strange Land*, by Robert A. Heinlein. The story focuses on one Valentine Michael Smith, the son of astronauts who had been sent to Mars, who was raised by Martians and then, some twenty years later, rescued and brought to Earth. To the great puzzlement of those who care for him, "Smith is not a *man*. He is an intelligent creature with the genes in the ancestry of man, but he is not a man. He is more Martian than a man" (Heinlein 1961, 20). One of his odd features is his ability to "grok," a kind of psychic power whereby one senses and blends with the being and thoughts of another, becoming as one with the other. This expresses the oneness or essential unity of life. Because Mars is so arid, the most meaningful ceremony for Smith is that of sharing a glass of water—whereby the participants become "water brothers." After initial difficulties here on Earth, Smith is allowed to roam freely. He encounters the Fosterite Church of the New Revelation, a megachurch whose members delight in practices normally forbidden or frowned upon—gambling, drinking, and (above all) lots and lots of free-ranging sex, especially among those in the positions of high authority. Smith founds his own church, the "Church of All Worlds," which incorporated Martian practices (especially those involving psychic ability) with those of the Fosterites, and was dedicated to rising above suffering and to spreading love and harmony. Inevitably this leads to conflict with the Fosterites, who consider Smith a heretic, and he is ultimately killed by them, refusing (in a way obviously intended to echo the

submission of an earlier religion-founder) to protect himself. His
followers (again echoing that earlier religion-founder) cook and
eat his body, and Smith ends the novel as a divine being with work
to do righting the physical world and its denizens.

Captivated by this story, Zell and his friends entered into Heinlein's fictional world. Unlike most undergraduates of the 1960s
who indulged in such protracted, adolescent fantasizing and then
went on to find fulfillment as university professors or stockbrokers,
Tim Zell and friends proved to be made of sterner stuff. They
founded a religion based on their play acting, naturally enough
known as the "Church of All Worlds," that was in some sense
dedicated to the organic unity or oneness of all life. The water
ceremony—sharing a drink from the same vessel—was especially
important. Zell—who is fundamentally indifferent to lines between fact and fiction and views Heinlein as important precisely
because he taps into universal themes (Plato-inspired Jungian
archetypes)—still speaks of this practice in almost pre-Platonic
terms as a "sacrament." Water has a mystical dimension, signify-
ing the interconnectedness and relatedness of all being. It is "the
universal essence of life everywhere" (OZ).

So much for doctrine. As the great evangelist from Tarsus
showed, to found a religion you also need organizing abilities.
Combining his passion with an open and engaging personality,
Zell showed a truly Pauline skill at promoting his ideas. One of his
first moves, having founded his church (of which he is Primate),
was to gain recognition for his church from the Internal Revenue
Service and consequent tax exemptions on its activities. He has
lectured extensively and published magazines (notably the leading Pagan organ, *Green Egg*). Although he has long been married
to Morning Glory (formerly Diana Moore of California), in the
spirit of *Stranger in a Strange Land*, he has been an enthusiastic practitioner of "polyamory" (consensual and responsible non-
monogamy), which at times seems to be almost as difficult to man-
age as celibacy (although probably more fun) and is intended to
express the oneness of all being ("One relates to one's partner as
an avatar of the divinity"). Chemistry also played a role, however

("the pill liberated us"). As we might expect from one who "got into drama" when young, Zell is enthralled with various kinds of elaborate rituals, preferably performed "sky-clad" (stark naked) to emphasize that coverings, be they clothes or myths, are not important, for what counts are the underlying unifying truths or foundations. "As Cicero said: '*Omnia vivunt, omnia inter se conexa.*' Everything is alive; everything is interconnected. This is what I consider the core of the Ancient Wisdom" (OZ).

Certifying Zell's place within the Pagan movement was a complete commitment to the environment and to Earth. On September 6, 1970, Tim Zell had a "profound Vision" that confirmed in him the belief that the entire Earth is one integrated living organism. Using the term *Terrabios* for what Lovelock was to call *Gaia* (Zell and his followers later adopted this term, often preferring the alternative *Gaea*), he called his sermon for the next service "Thea-Genesis: The Birth of the Goddess." He asserted that, since all life comes from an original cell, all life is therefore not just related but is one. "It is a biological fact (not a theory, not an opinion) that ALL LIFE ON EARTH COMPRISES ONE SINGLE LIVING ORGANISM! Literally, we are *all* 'One'" (Zell-Ravenheart 2009, 92). He goes on to say, "The blue whale and the redwood tree are not the largest living organisms on Earth; the ENTIRE PLANETARY BIOSPHERE is." Individual organisms are the cells of Terrabios. The deserts, the forests, the prairies, and the coral reefs (the "biomes") are the organs. "ALL the components of a biome are essential to its proper functioning, and each biome is essential to the proper functioning of Terrabios" (92).

As in the work of Lovelock and Margulis, the inorganic parts of the earth are not neglected; indeed, they are seen to be a vital part of the whole. Thus, for instance, "The rock and mineral foundation of our planet functions in the body of Terrabios much as the skeleton functions in the human body" (Zell-Ravenheart 2009, 93). Likewise, "The water of oceans, lakes and rivers that covers three-quarters of the surface of the globe functions homologously with the plasma in the blood of the human body, which incidentally has a composition very like the water in those primeval seas wherein life first appeared." The atmosphere functions for Terrabios much as it does for individuals, and the sun provides the energy for Terrabios as it does for single organisms. Moreover, the parts of the whole of Terrabios have the same relationship to the whole as do the parts of the individual body. Altering or removing parts has consequences for the whole. "You can't kill all the bison in North America, import rabbits to Australia, cut or burn off whole forests, or plow and plant the Great Plains with wheat without seriously disrupting the ecology. Remember the dust bowl? Australia's plague of rabbits? Mississippi basin floods? The present drought in the Southwestern U.S.? Terrabios is a SINGLE LIVING ORGANISM, and its parts are not to be removed, replaced, or rearranged" (93). Within the system of Terrabios, humans are the biome of awareness. Borrowing a term from the French Jesuit paleontologist Teilhard de Chardin, we are (humans taken collectively) the *noösphere*. Our function is "to act as the steward of the planetary ecology." In short, "Man's purpose in Terrabios, his responsibility, is to see that the whole organism functions at its highest potential and that none of its vital systems become disrupted or impaired" (93). "Blessed Be!"

This is unfamiliar territory for most of us, but, as predicted at the beginning of this chapter, not entirely new. "I could probably be said to be a neo-Platonist to a great extent—especially the more holistic, monistic elements" (OZ). Otter G'Zell, as our wizard was known for a while, mid-point in his transition from Tim Zell to Oberon Zell-Ravenheart, spotted at once the links between his world picture and that of Jim Lovelock, and he corresponded enthusiastically with the English scientist. And yet— groking, polyamory, Terrabios! Such ideas and practices are not found in the world of professional science. This is the world of religion, laced with pseudoscience. What a strange journey, from an inspiration born out of space science to practicing consensual and responsible non-monogamy. Has Gaia vanished into the aromatic fug of the Californian counterculture? It is obviously time to pick up the story again with our main protagonists.

7

GAIA REVISITED

We return now to the 1980s and to James Lovelock and Lynn Margulis. Let us take them in turn and explore how the notion of Gaia played out in their lives and thinking. I leave until the next chapter an overall assessment of their work and a resolution of the paradox that set us on our path: why the scientific community reacted so negatively to the Gaia hypothesis, whereas the public reaction was so positive.

JAMES LOVELOCK

When we left James Lovelock in the early 1980s, he was feeling a little battered psychologically. He had taken a fairly heavy drubbing from the biologists, especially Richard Dawkins and Ford Doolittle. His response was partly petulant. He tended to go off on a tangent, pointing out that he was an independent scientist living by his wits and that he alone, therefore, had the scope and opportunity and (by implication) the courage to think outside the circle or, more precisely, to enlarge the circle in order to understand Earth through many different sciences, including geology, chemistry, and physics as well as (by implication) inward-looking biology. Institutions, grant-seeking, and, above all, narrow-minded reviews kill the creative spirit. "To cap it all in recent years the 'purity' of science is ever more closely guarded by a self-imposed inquisition called the peer review. This well-meaning but narrow-minded nanny of an institution ensures that scientists work according to conventional wisdom and not as curiosity or inspiration moves them. Lacking freedom, they are in danger of succumbing to a finicky gentility or of becoming, like mediaeval theologians, the creatures of dogma" (Lovelock 1988, xiv). Biologists are especially to blame for their exclusive focus on their own domains of expertise, and, while they are good at handing out criticisms, they "may not have noticed the extent of their own errors" (33). The reception Gaia got from the life scientists still rankles: "Biologists, biologists, biologists hated it, loathed it, God knows why" (JL). Lovelock does have his suspicions, however: "The great bulk of biologists become biologists because they can't handle the mathematics" (JL).

Yet Lovelock was not about to back off. He suggested that we should start using less provocative names, such as *geophysiology* for the science that covered the insights he was promoting, but still the old language and thinking prevailed. "The Gaia hypothesis supposes the Earth to be alive, and considers what evidence there is for and against the supposition" (Lovelock 1988, 8). Provocatively, he wrote a popular book called *Healing Gaia: Practical Medicine for the Planet* (1991). The theme was that Gaia was an organism, in trouble today because of humans and in need of planetary health care. "Gaia is the Earth seen as a single physiological system, an entity that is alive at least to the extent that, like other living organisms, its chemistry and temperature are self-regulated at a state favourable for life" (Lovelock 1991, 11). He does not use the term *paradigm*, but the intimations are that something major is going on and he, Jim Lovelock, man of independent spirit, is at the heart of it all. "The progress of good science is slow and unpredictable and all too often waits upon the appearance of a key thought in the mind of a genius. The mere employment of a hundred new and brightly polished doctors of philosophy from great universities to tackle the problem of global change is most unlikely to achieve anything other than provide them with secure and comfortable employment" (15). Worth mentioning is that back when they were collaborating, Margulis seems to have thought that they were making a paradigm (Clarke 2012), and since then Lovelock's followers have not hesitated to talk in Kuhnian language, suggesting that the "debate over Gaia has all the hallmarks of different scientific communities grappling with a paradigm shift" (Lenton 2002, 419). In other words, ask not for whom the bell tolls; it tolls for everyone else.

What does make Lovelock stand out is that he is a very good scientist—he's right, he is a genius—and, if only subconsciously, he realized that he had to respond as a scientist to the attacks. He persisted in his earlier attempts to change the terms or focus of the debate. The chief criticism of the biologists was that Earth could not possibly be an organism, because organisms are produced by natural selection, and not even metaphorically could one think of Earth as being produced by natural selection. In *Healing Gaia* Lovelock wrote, "A neo-Darwinian biologist will define a living organism as one able to reproduce and correct the errors of reproduction through natural selection among its progeny" (Lovelock 1991, 29). From this follows the objection: "Richard Dawkins is a passionate advocate of reductionist science, and he and many of his followers oppose and ridicule Gaia theory. Gaia cannot reproduce, they say, and therefore cannot evolve in competition with other planets. Therefore it cannot be alive" (29). But Lovelock questions the right of biologists to assume that this is the defining characteristic of a living organism. "It is true that Gaia is not alive like you or me. It has no sense of purpose, it cannot move by its own will, or make love. But then neither can many bacteria. Are these not alive? And what about grandmothers, they cannot reproduce; neither can Lombardy poplar trees, all of which are male. Or, on a larger scale, what about whole ecosystems, such as forests? Are all of these, which we thought were living, to be pronounced dead?" (29–31). In any case, "Life is a planetary-scale phenomenon. On this scale it is near immortal and has no need to reproduce" (Lovelock 1988, 63).

In Lovelock's candid opinion, despite the confident assumptions of the neo-Darwinians, a search of the literature reveals that there is a bit of a black hole on the subject of what defines life. "Even scientists, who are notorious for their indecent curiosity, shy away from defining life. All branches of formal biological science seem to avoid the question" (Lovelock 1988, 16). Lovelock therefore felt justified in trying his hand at a definition. He concluded that life itself was to be defined not in terms of natural selection but in terms of the notion emphasized by W. B. Cannon, the internal balance of an organism: homeostasis. Dawkins saw an adaptation like homeostasis as something consequent to the working of natural selection. Lovelock stressed (surely showing his training as an applied chemist rather than an evolutionary biologist) what had always been his position, that it is homeostasis in itself that is all-important, the very defining essence of life. "Life is social. It exists in communities and collectives. There is a useful word in physics to describe the properties of collections: *colligative*. It is needed because there is no way to express or measure the temperature or the pressure of the single molecule. Temperature and pressure say the physicist, are the colligative properties of a sensible collection of molecules." Then there is an apparently holistic conclusion. "All collections of living things show properties unexpected from a knowledge of a single one of them. We, and some other animals, keep a constant temperature whatever the temperature of our surroundings. This fact could never been deduced from the observations of a single cell from a human being" (Lovelock 1988, 18).

Lovelock felt comfortable with being quite explicit: "The concept of planetary medicine implies the existence of a planetary body that is in some way alive, and can experience both health and disease. . . . I often describe the planetary ecosystem, Gaia, as alive, because it behaves like a living organism to the extent that temperature and chemical composition are actively kept constant in the face of perturbations" (Lovelock 1991, 6). Lovelock stressed that Earth was not an organism in the same sense as humans are, for "when I talk of a living planet, I am not thinking in an animistic way, of a planet with sentience, or of rocks that can move by their own volition and purpose. I think of anything the Earth may do, such as regulating the climate, as automatic, not through an act of will, and all of it within the strict bounds of science" (31).

However, he emphasized that the concept of an organism extends beyond the animals and plants on planet Earth, and encompasses the rocks, oceans, and everything else. "The biota and the biosphere taken together form part but not all of Gaia. Just as the shell is part of a snail, so the rocks, the air, and the oceans are part of Gaia" (Lovelock 1988, 19).

DAISYWORLD

Hurt and defiant Lovelock may have been. Inventive, as always, Lovelock certainly was. As we saw in chapter 1, a major plank in Lovelock's thesis about Earth being in homeostasis is how, despite the fact that the sun has become much hotter over the years, life keeps our planet's temperature relatively constant. The early papers (e.g., Margulis and Lovelock 1974) offered a rich diet of suggestions as to how this happens, including for instance the greenhouse effect. As the sun heated up and sent more solar radiation our way, it seems that the amount of carbon dioxide in the atmosphere was reduced. This meant that (at least until recently) the greenhouse effect was reduced—less heat was trapped by the atmosphere, and the new amount of heat coming in was balanced by the extra heat leaving. In a way, this seems like a one-off sort of thing. Serendipitously, something kicked in to keep temperatures constant. It doesn't seem as if there is a mechanism focused on keeping the system in equilibrium. But is this truly so? How could a system stay in equilibrium or regain it if lost? For this—for the kind of homeostasis that Cannon was talking about, we need to think in terms of feedback systems. There are two kinds of feedback systems: positive and negative. Positive feedback simply increases or magnifies the disturbance; negative feedback, which is of interest here, tamps it down. Something in the system is triggered that pushes the system toward its original state. Lovelock, whose real genius lies in his skill at gadget and machine building, noted that a classic example of negative feedback is the eighteenth-century engineering marvel that we discussed in chapter 3, the steam engine governor (Lovelock 1991, 60). As the steam engine heats up and provides greater force, a shaft driven by the engine rotates faster and faster. Two heavy metal balls attached to the shaft by hinged arms rotate with the shaft. As they speed up, they get flung into a wider orbit and rise, thus opening a valve through which steam can escape. This reduces pressure, and the engine slows down. This goes on indefinitely. Every time the engine speeds up, it causes the event that slows it down. This is negative feedback.

Negative feedback is well known in physiology. We have already mentioned the prime example of keeping the body at a constant temperature through sweating and shivering. Can we expect to find similar feedback systems controlling planet Earth and keeping it in equilibrium? Here the biological critic Paul Ehrlich simply parted ways with Lovelock. "There has been an enormous impact on the physical character of the planet by the organisms of the planet. Obviously, there also are homeostatic aspects to the system. . . . There clearly are many interesting negative feedback loops, as well. Nonetheless, I find it very hard to believe that the physical Earth in some direct or even indirect sense is evolving to make life comfortable for the organisms on it." For Ehrlich and others, there is something too end-directed, too teleological, about this kind of thinking. And there are no models or mechanisms, just analogies between Earth and organisms. "That is why in the context of Gaia, I find myself taking a reductionistic position; the idea that life evolves in a way to make the planet more hospitable for itself collapses for want of a mechanism" (Ehrlich 1991, 21).

Lovelock realized that he had to resolve this problem. He admitted that he was "dead wrong" not to have tackled it from the start (JL). He had to explore feedback systems and show how (without getting illicitly teleological) equilibrium could be achieved and maintained. One senses from some of the later discussions (e.g., EE) that the greenhouse effect was considered as the really important controlling factor; but, as just pointed out, there was no specified reason for the control. The critics really did have a point. It was as if one had all of these mechanisms for achieving balance, and behind them a kind of life-force was pulling the strings. A mechanism had to be found to show that negative feedback rather than positive feedback was at work and that this led to equilibrium. In Aristotelian terms, the Darwinians argued that the only permissible sense of final cause is one connected to natural selection. Lovelock had to show that his proximate-cause mechanisms could likewise lead to some kind of desirable end, some kind of final cause.

In pursuit of this goal, what Lovelock did next was deeply revealing. When you encounter a problem, the obvious first move is to see if there is an established and respected body of knowledge—an area or school of science—that would help and with which you could ally yourself. And there actually was such a school based on the work of G. Evelyn Hutchinson, mentioned earlier, and his students and followers, most notably the brothers Howard T. and Eugene P. Odum. The accompanying diagram on the carbon cycle from Hutchinson's classic paper, "Circular Causal Systems in Ecology," published in 1948, clearly shows that the Yale ecologist was totally committed to feedback systems as a tool of understanding and also fully aware that a complete grasp of the problem involved both the organic and the inorganic (fig. 17). Even more pertinently, in a presidential address Eugene Odum explicitly likened ecosystems to organisms and described them as growing until they reached some kind of equilibrium or homeostasis (his words). "As viewed here, ecological succession involves the development of ecosystems; it has many parallels in the developmental biology of organisms, and also in the development of human society. The ecosystem, or ecological system, is considered to be a unit of biological organization made up of all of the organisms in a given area (that is, "community") interacting with the physical environment so that a flow of energy leads to a characteristic trophic structure and material cycles within the system" (Odum 1969, 262).

For Lovelock, this should have been manna from heaven, especially because Eugene Odum (1989) actually rather liked the Gaia idea. Yet Lovelock simply did not draw on this work. Why? It was not through ignorance. Although he did not know the Odums' work, Lovelock knew of Hutchinson's work, and referenced it often, especially in his early papers with Margulis. They even went

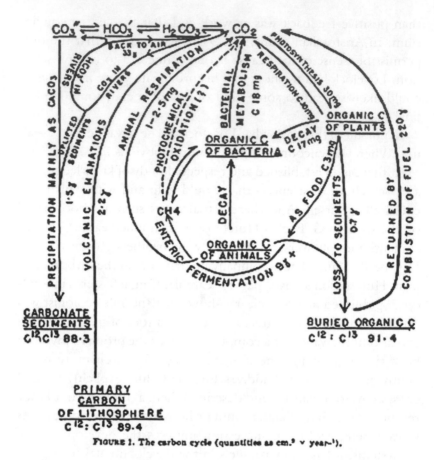

FIGURE 1. The carbon cycle (quantities as cm.⁹ ᵛ year⁻¹).

Figure 17. G. Evelyn Hutchinson's carbon cycle. (From G. E. Hutchinson, "Circular Causal Systems in Ecology," *Annals of the New York Academy of Science* 50 [1948]: 221–46. 1948, figure 1. © The New York Academy of Sciences.)

so far as to say, "We are also indebted to Prof. G. E. Hutchinson, whose life work has surmised rather than explicitly acknowledged the existence of Gaia" (Lovelock and Margulis 1974a, 103). Part of the reason for Lovelock's refusal to draw on Hutchinson's work was surely his contempt for biologists, noted above, which was no doubt particularly strong at the beginning of the 1980s, thanks to the drubbing he was getting. This was intensified by what seems to have been a rather disastrous meeting with Hutchinson, when a third party who was hostile to Gaia dominated the conversation (Lovelock 2000, 263). We might also add that, in light of the severe criticism of his hypothesis, it would have been altogether too saintly of Lovelock to simply hand over to others the credit for priority and a superior approach.

Basically, though, although he had spent much time in medical research, Lovelock was not a biologist, didn't think like one, and wasn't about to start now. Instead, to use his own language, he thought like a computer nerd. He had been building his own computers from kits since the 1960s and had spent time and money to stay up to date with the technology. He sought and found an idea that lent itself to computer modeling—the idea (developed with a student, Andrew Watson) of Daisyworld (Watson and Lovelock 1983). The concept is quite simple, although over the years (mostly because it can be readily modeled on computers), it has become more complex and sophisticated. Think of a planet like Earth going around a sun like our sun. Unlike our planet, however, it has only two species of organism, light-colored (white) daisies and dark-colored (black) daisies. They grow between (let us say) five degrees Celsius and forty degrees Celsius, with the best growth somewhere in the middle at around twenty-two degrees Celsius. Dark daisies readily absorb the sunlight and hence are prone to heat up to a temperature higher than their surroundings. Light daisies reflect more of the sunlight, so they tend to be cooler than their surroundings. As the sun heats the planet, at some point the ambient temperature reaches the five-degree (planet surface) level. Both kinds of daisies can now start growing, but the dark ones will do better than the light ones because they heat up more and hence reach the optimal growing temperature sooner. They will spread rapidly and dominate the planet's surface—until, that is, their numbers (and the ever-increasing heat from the sun) take them beyond the optimal conditions. Now the light daisies are favored; their numbers increase and reflect more and more sunlight, and this begins to bring the temperature of the whole planetary system down. Negative feedback leads to a balance that persists

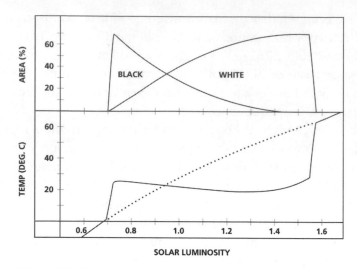

SOLAR LUMINOSITY

Figure 18. Diagram showing the dynamics of Daisyworld. The relative proportions of white and black daisies can lead to a stable temperature on their planet. (From A. J. Watson and J. E. Lovelock, "Biological homeostasis of the global environment: The parable of Daisyworld," *Tellus, Series B: Chemical and Physical Meteorology* 35 [1983]: 284–89.)

despite the increasing heat from the sun. Finally, however, the heat from the sun increases so much that no daisies can grow, and the whole system breaks down (see fig. 18).

As the star increases its heat output, . . . the dark daisies decline and the population of light-coloured daisies spreads. The temperature remains close to that preferred by daisies. Like the electric iron, the system tends to regulate on the hot side of optimum when the starlight is weak at the beginning, and on the cool side when the output of the star is great, later on. Eventually in the evolution of the star, its output grows so great that even a planet-wide cover of the heat-reflecting light-coloured daisies is insufficient to keep a tolerable climate for daisies; the system suddenly and catastrophically fails, and Daisyworld dies. (Lovelock 1991, 67)

Notice that, although natural selection is not working to promote the planet as such—nor is the planet itself a product of selection—the model shows natural selection working at the level of the individual daisy. Black daisies are at a selective advantage when the heat from the sun is relatively low, and white daises have the advantage when it is high; between the extremes, the two kinds of daisies are promoting their own interests (having offspring) and balance each other out. Obviously, this is just a model, and now some empirical spade work is required to show how it or something like it applies to Earth, but the principle has been demonstrated. Lovelock had produced a mechanism that conformed to orthodox Darwinism based on individual selection in an attempt to rid his theory of any hint of teleology. No one is planning anything, nor are there unseen forces driving things to needed ends. It is all a matter of machines in motion.

GAIA: PRO AND CON

In the 1980s, the literature on Gaia grew exponentially (Schneider and Boston 1991; see their references). Despite Lovelock's defense of Gaia, there continued to be much criticism. A common empirical worry was that Earth has gone through some pretty major changes; for instance, the arrival of large quantities of atmospheric oxygen (after the development of photosynthesis) brought on a major change in the overall nature of life on Earth. This argued against homeostasis. The response was that homeostasis does not preclude change. "Homeostasis in living systems is not a permanent, fixed state of constancy; it is a dynamic state of constancy" (Lovelock 1991, 141). Only those ignorant of the nature of machines could make such a charge. The gyroscope used in a ship's autopilot wobbles and fails if it is simply allowed to run down, but if the ship is redirected, then a new state of stability is soon achieved. The process of going through such rapid changes, added Lovelock helpfully, is known as "homeorhesis."

At a more conceptual level, the objections came most persistently and effectively from the earth scientist James Kirchner, who had the advantage of philosophical training that he used to good effect (Kirchner 1989; 1990; 1991; 2002; 2003). Lovelock seems to have thought the criticisms unfair, but perhaps this confuses vindictive intent with honest disagreement, for while Kirchner was always blunt, he was never less than courteous. Kirchner's major complaint was that it was difficult, if not impossible, to assess the truth value of Gaia—to subject it to any kind of meaningful test—because the discussion was too loose. It concealed or conflated many different claims, some of which were true and never doubted, and some of which were almost certainly not true and hence were completely doubted. Of course, Gaia may be a metaphor and nothing more, something intended to stir the creative juices but not to be taken literally. Kirchner had no objection to this, but he argued that if it was no more than a metaphor, then one should be wary of making truth claims. Almost by definition these are ruled out. It is not really meant to be a scientific proposition in the first ·place. "Gaia is crippled by its great generality; it searches for a simple capsule description of the role of life on Earth," Kirchner wrote. "Gaia may be a grand vision, but it is not the kind of vision that can be scientifically validated" (Kirchner 1989, 233–34).

Systematically, if not always patiently, Lovelock kept responding to his critics, tweaking here, expounding there. He realized that, as a general rule, the worst option for scientists faced with criticism, especially niggling philosophical criticism, is to criticize the critics. Far better to ignore it and make the critics eat crow by showing that, judged by standards of explanatory effectiveness and predictive ability and fertility, one's work does indeed lead to good quality science. In addition to his theoretical defense in terms of the Daisyworld models, Lovelock got back into the discussion with explicit examples intended to show the importance of Gaia-based thinking. In a major article in *Nature* in 1987, Lovelock, collaborating with other scientists, argued that sulfur is part of a significant feedback loop controlling Earth's temperature (Charlson et al. 1987). Clouds don't just happen. You need something to "seed" them, particles in the air around which water droplets

can form. Among the many possibilities for such "cloud condensation nucleics" (CCNs)—dust, carbon in various forms, and so
forth—one candidate is sulfur, particularly in the oxidized form
of sulfur dioxide (SO_2) gas. Moreover, it appears to be significant, particularly over oceans. Other possibilities can be eliminated
or are relatively trivial. Sea salt particles, for instance, tend to be
too small to be effective. Nitrate particles, conversely, seem to be
too big. Carbon particles are too rare. However, there does seem
to be enough sulfur around to do the job.

The question then becomes one of origins. Where does this
sulfur come from? One possibility is that the source is geological.
SO_2 and H_2S (hydrogen sulfide) spew out from volcanoes. But this
source seems capable of producing at most only 10–20% of what
we find. The search therefore turns toward the organic—most obviously to the algae that bloom in the sea. Could planktonic algae
be the source of the sulfur one finds in the atmosphere? Indeed,
this does seem so. The key is dimethylsulfide (DMS), a substance
excreted by algae. Although the exact reason for the production of
DMS is not known, it is a product of another substance, dimethyl-
sulfoniumpropionate (DMSP), that is needed for osmoregulation
(keeping the internal salt content at a livable level). Whatever is
going on, an important point is that apparently the warmest parts
of the oceans produce the greatest quantities of DMS. There are
several reasons for this. One seems to be related to the greater
ventilation over warmer areas. Another is that the species of algae found in warmer waters produce more DMS than do those in
colder waters. "It appears that some algal groups, such as the coc-
colithophorids, which are most abundant in tropical, oligotrophic
waters, have the highest rate of DMS excretion per unit mass"
(Charlson et al. 1987, 656).

We are now just about ready to postulate a feedback loop demonstrating that organisms regulate the planetary system so as
to keep conditions constant despite potentially varying external
conditions. The sun warms the ocean surface, which triggers the
growth of algae that produce more and more DMS. This rises, gets
oxidized into SO_2, and promotes cloud formation. The clouds re-

flect the sun's rays, and the surface of the ocean starts to cool down. This triggers reduced algal growth, which results in reduced DMS production. So we end up with a balanced situation—homeostasis in action: "The clouds [serve] as do white daisies in the 'Daisyworld' model of Gaian climate regulation" (Charlson et al. 1987, 661).

We might even conceive of a bigger loop; sulfur drifts over to the land, promoting growth and consequent weathering there. The sulfur and other nutrients necessary for algal growth then return to the sea. This may seem a bit farfetched. "Why should an algal community of the ocean make the extravagant altruistic gesture of producing DMS for the benefit of, among other things, elephants and giraffes?" (Charlson et al. 1987, 660). Perhaps, suggest the authors, it is something left over from the original function of salt-content regulation. "We begin to see a possible geophysiological link between the local self-interest of salt-stress prevention and the global sulphur cycle. The accidental by-product of DMSP production is its decomposition product DMS. This compound or its aerosol oxidation products will move inland from the shore and deposit sulphur over the land surface downwind of the ocean. The land tends to be depleted of sulphur and the supply of this nutrient element from the ocean would increase productivity and the rate of weathering and so would lead to a return flow of nutrients to the ocean ecosystems" (fig. 19). In short: "What seems a naive altruism is in fact an unconscious self-interest. Sulphur from DMS can travel farther than the sea-salt aerosol because several steps are involved in the conversion of gaseous DMS to aerosol sulphate; also the resulting aerosol particles are much smaller and so have much longer lifetimes" (660).

The Gaia hypothesis seems to be vindicated. Lovelock felt that he was getting into the mainstream. Twenty years after the original paper appeared, he was still citing it with contented enthusiasm. "Several years later in 1986, while collaborating with colleagues in Seattle, we made the awesome discovery that DMS from ocean algae was connected with the formation of clouds and with climate. We were moved to catch a glimpse of one of Gaia's

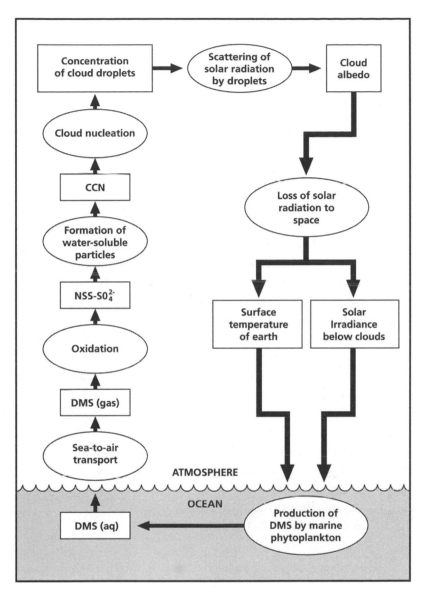

Figure 19. The DMS cycle, somewhat simplified to stress the feedback circle. (From Charlson et al. 1987, fig. 2, 659.)

climate-regulation mechanisms, and we were indebted to the climate-science community who took us seriously enough to award to the four of us . . . their Norbert Gerbier Prize in 1988" (Lovelock 2006, 23). This prize is awarded by the World Meteorological Organization, a branch of the United Nations. The awarding committee apparently did not include Kenneth Caldeira, an ardent Darwinian evolutionist, who did not agree that Lovelock and company were making a Darwinian case. Production of DMS for the benefit of others just does not happen in the world of the selfish gene. Any population altruistically making DMS other than for its own ends is wide open to invasion by organisms that do no such thing. Nor is there reason to think that this might be one of those rare occasions where group selection is a possibility. For this to occur, the altruists would have to be grouping with other altruists in order collectively to overwhelm the non-altruists. But such conditions don't seem to be present in the algae case. Nor does it seem that individual selection can do the job here. There is no reason to think that DMS production might also benefit the producer. Lovelock and his associates "suggest that the DMS emitters might be at an advantage because cloud formation with rainfall would return nitrogen to the ocean and also serve as a sunshade. But if DMS emission increases cloud albedo and diminishes the solar flux of the surface waters, globally, less energy will go into latent heat of evaporation and thus rainfall might actually diminish. Locally, increased CCN number density would lead to smaller cloud droplet size and a diminished likelihood of rainfall, increasing the probability that the clouds would be advected [carried horizontally by the air] to other regions" (Caldeira 1989, 733).

HAMILTON AND GAIA

With these arguments for and against Gaia, we have reached the end of the old century and are moving into the beginning of the new. To round off this stage of the discussion, let us turn to the surprising engagement with Gaia by the man rightfully regarded as the most creative Darwinian of the past half century, Oxford evolutionary biologist William D. Hamilton. I call it surprising because it was Hamilton who, as noted in an earlier chapter, was most responsible for the recognition that selection at the individual (rather than the group) level is what really counts. He was the man behind Richard Dawkins's "selfish gene" perspective on evolutionary mechanisms. And yet, although individual selection is the most significant challenge for the Gaia supporter, perhaps Hamilton's (far from purely negative) engagement with Gaia was not so surprising because he had developed a well-justified reputation for applying his kind of thinking in new and unexpectedly fruitful ways in his studies, for example, of hymenopteran (ants, bees and wasps) sterility, competition among siblings, the evolution of sex as a defense against parasites.

A series of Gaia conferences were held in the 1990s in Oxford, and, following the second (in 1996), Hamilton wrote a long, friendly, but very critical letter to Lovelock, expressing just about every concern that a scientist, especially an evolutionary biologist, could have about Gaia. The original Dawkins objection was up front. Obviously, Lovelock does "regard Gaia as a superorganism," and moreover thinks "the 'parallel' of the homeostasis of social insect colonies as a legitimate one, such that investigation of the homeostasis of the one might illuminate the homeostasis of the other." But this cannot be. For Hamilton, it is quite obvious "that *gaia can't be a superorganism in the same way that social insect colonies are superorganisms.* The latter have evolved their self-regulating properties through millions of years of natural selection—more stable, more homeostatic variants tending to replace less stable ones" (his emphasis). Hamilton continued,

You take a word, "organism," which has acquired an accepted meaning among Darwinists as a unit of a system subject to natural selection, and you apply [it] to a system that can't possibly be subject to n.s. [natural selection]. Because there is only one of them! Obviously that is equivalent to asking everyone to change their concept of an organism. This may be what you want, but you should not be surprised at resis-

tance! I think you would get further in attracting interest if
you were to invent a new term—well, I suppose "Gaia" is actually that—and explicitly deny that you are saying the planet
is an organism. Just calling it a "super" organism doesn't
help here because that term is already in use by Darwinists
for the objects like social insect colonies where n.s. is still
in play. (Unpublished letter from W. D. Hamilton to J. E.
Lovelock, January 29, 1997)

Perhaps somewhat unkindly, Hamilton likened Lovelock's use of
organism in the absence of natural selection as akin to someone
asking for grant money in the field of geochemistry for such a
nonsense project as "the quantum status of the earth-moon super-
atom system."

Of course, Lovelock could respond that he does in fact rely on
natural selection, as in the Daisyworld model. But such thinking
got short shrift in Hamilton's letter: "You are right in saying you
have n.s. in your model, but it is of a very simplistic type so far
and to my mind is very far from showing that increasing stability
is an inevitable outcome in a global system of planet+biota." What
we need is a principle showing how natural selection promotes
planetary stability. In any case, Hamilton went on, "For me, it is
actually much stronger evidence if Gaia-like properties emerge in a
model that was conceived completely independently of the idea it
turns out to illuminate." Ultimately, of course, we still find teleology lurking here. We just don't seem to have grasped the reason
why everything, including feedback loops, should add up to stability. For instance, the idea (suggested by one Gaia enthusiast)
that "peatlands send organic iron to organisms in the sea and they
induce the sea to send sulphur back" struck Hamilton as almost
requiring "treaties between Neptune and Zeus, a Gaian Interpol,
conventions about bills of lading for chemical transport by air and
water, and so on, to make it work." And finally, rubbing salt into
the wound, Hamilton sniffed that name changes were all very well,
but he was frankly indifferent:

I have little objection to the newly named discipline "geo-physiology," but that is possibly because I'm not a physiologist and don't feel any great involvement with the word. . . . But, much as I feel about physiology in the normal sense, . . . it is all rather mechanistic and somewhat descriptive science for my taste. It is important science, and I don't doubt that taking into account the feedbacks from life and your various discoveries of new links in the system can give us better predictions. However, this is all answering different sorts of questions from those of the evolutionary biology.

Those are the questions that Hamilton wanted answered, and "to make Gaia seem life-*like* enough for my interest to be ensnared, this matter of a new principle predicting increasing stability is really crucial—at least to me."

Yet something about Gaia caught Hamilton's interest, because he was soon collaborating with Tim Lenton, a young supporter of Lovelock, and studying one of those feedback loops that figure so prominently in the discussion. Start with the studies showing that iron is a significant factor in algal growth. The main source is from water welling up from the deep, but dust is also a factor, and during ice ages (possibly thanks to the drying effect of colder temperatures and also to an increase in winds, not to mention fewer plants to anchor the soil) dust may become relatively much more important because it gets blown out to sea from the land. During an ice age, therefore, the iron supply for algae rises, which fosters greater populations of algae and subsequent release of more DMS. In turn, we get more clouds and more solar reflection, and hence more cooling. This positive feedback—cold, dust, algae, DMS, cloud, colder—helps to explain the intensity of the ice age, something evidently linked to rises in sulfur content in the air (Watson and Liss 1998). One matter of interest, especially to an evolutionary biologist, is whether anything in the cycle benefits the algae. Is it all a matter of by-products, or is natural selection actively involved, particularly in the production of DMS? This is an especially significant question. If DMS is just a by-product, and if the feedback cycle is thus pure chance from the biological view-point—even if the algae benefit from its cooling effects—then this neither confirms nor refutes evolution through natural selection. In the light of the criticism above, however, we probably need some argument to show that releasing DMS does not cost the algae anything in terms of reproductive success. But if we can show that natural selection is positively integrated into the story, then, whatever the status of the Gaia hypothesis, we are moving empiri-cally toward defusing the tension between this approach to Earth science and conventional Darwinian evolutionary theory.

Hamilton seized on the need for dispersal. Organisms are al-ways looking for new fields to conquer—preferably new fields far enough from home that they will not encounter the competitors of their original environment. "Dispersal is extremely important to life, indeed for self or progeny it can be considered an organism's third priority after survival and reproduction. It remains a crucial necessity even if growth conditions at the point of landing are never better than at the point of takeoff, although if they are some-times better, the incentive is of course all the stronger" (Hamilton and Lenton 2005, 275). The important point however, of which Hamilton was well aware, is that one must get around any sugges-tion of group selection. "The precursor to DMS in marine algal cells, dimethylsulfoniumproprionate (DMSP), is plausibly evolved as an osmolyte buffering plankton cells against salt-concentration changes and sometimes ice damage. However, the step of claim-ing that an initial side-effect of DMSP synthesis—namely DMS production—was adaptively seized on in some way to regulate the climate of the planet, is controversial." Groups simply do not prevail over individuals. "Indeed, in the claim's simplest form, a *Darwinian* mechanism for additional expenditure for such an end can be excluded: a benefit that increases the welfare of an entire group considered in isolation does not increase the frequency of the causative element. Worse, if creating the benefit had a cost, the gene causing the group benefit definitely declines" (274).

Selection must be at the level of the individual or (ultimately the same thing) at the level of close relatives, including clones. The latter of course are all genetically identical, so what happens to one is something of benefit (or loss) to all. Perhaps, suggested Hamilton, the production of DMS really is adaptive. Caldeira, the critic of Lovelock's paper on sulfur and feedback, made the point that grazing and cell death can cause DMS release. "The rate of DMS release by phytoplankton increases greatly when the phytoplankton are subjected to grazing by zooplankton, indicating that much DMS release is a result of cell rupture and death. DMS gas production is 7–26 times higher during the senescence phase of a phytoplankton bloom, when cellular integrity is lost for many phytoplankters, than during the growth phase" (Caldeira 1989, 734). This Darwinian critic used this point to refute the adaptive value of DMS release. What Hamilton pointed out was that the DMS thus produced cannot help the individual producing it, but it can help the clones and so could be favored by selection. The DMS follows the path shown in figure 19, forming clouds. But why would the clouds help the clones down below? Because many of the clones are no longer down below! There is strong evidence that algae, which are after all very small, can escape from the sea by means of bubbles forming on the surface, which are whipped up by the winds and soar aloft. Clouds add to the turbulence of the air, and this can enhance the lifting process. "Especially on sunny days with their high midday peak of radicals in the air, just when local convection is strongest, there seems a good chance for some situations in which convection due to algal DMS emissions generate a local increase of wind speed in a matter of hours." Selection can now get involved. "If white tops are augmented or initiated by this increase, then the take-off process already described can potentially pay back to DMS-emitting algae at the individual or clonal patch level of selection (or to individuals via inclusive fitness) an extra possibility for causative genes to become airborne, and to disperse rapidly away" (Hamilton and Lenton 2005, 274–75). Once aloft, algae need protection, especially from such things as ultraviolet radiation, and clouds can provide such protection. Most important, clouds are a means of transport from the original place to new grounds (or seas, rather). In other words, clouds are "extended phenotypes" in the language of Richard Dawkins, and their formation is adaptively favored by the algae. Even if the actual algae creating the possibility of clouds are dying or eaten, their genetically identical relatives benefit.

Was the world's most creative Darwinian now a Gaia enthusiast? This was certainly the excited conclusion of science writer Lynn Hunt in the *New Scientist*: "By explaining why microbes produce clouds, he [Hamilton] has also formulated the first biologically credible mechanism for Gaia—the theory that Earth acts like a superorganism, with all its biological and physical systems cooperating to keep it healthy." She continued as follows.

Hamilton is the first to admit that it all seems a little extraordinary. These ideas have convinced him that Gaia may be a real biological phenomenon, but how will they fare among his peers? "We are expecting a very defensive response from professional meteorologists, algologists and others working in this area." And he anticipates the response with some excitement. "This is a new view of the properties of microbes, a new explanation for why they do what they do."

And perhaps it is a new perspective on the climate system. We can begin to see how life on Earth is involved in regulating climate, not just passively, but actively. "This marks a turning point in the Gaia theory," says Lovelock. "Biologists are beginning to take the ideas seriously." (Hunt 1998, 33)

This provoked an immediate response (in a letter to the editor) from Hamilton. He acknowledged that he and Lenton had done some interesting science and had demonstrated what could be an important loop, but insisted that this was a long way from Gaia. "The natural selection mechanism we suggest could, it is true, be a component of a thermostat, but no one has yet shown how the thermostat could be adjusted appropriately. Indeed, it is not even clear if this thermostat is wired the right way: it could 'click' on and stay on until the world freezes for all that our study shows. It may be that a thermostat is there, but the evolutionary (or other) process leading to it needs to be explained" (Hamilton 1998). Hamilton went on to repeat points made in his private letter to Lovelock (January 29, 1997), namely, that without natural selection working on the system as a whole, he could see no reason to speak in terms of organisms. And if it is protested that Earth does indeed show homeostasis (not Hamilton's claim), then some reason must be sought, or perhaps it must be accepted that the appearance of homeostasis is just that, an appearance, and no more. "For the time being, however, it seems best to assume that the planet is subject to a set of out-of-control thermostats, and that human rationality, via science, is its main hope for homeostasis." In other words, group-selection thinking may have been eliminated from the system, but this only reveals more harshly the extent to which there is an underlying, unacceptable teleology built into the Gaia worldview. Lovelock (1998) was "disappointed" by this response. But for now we can leave him and his critics, and defer our analysis.

LYNN MARGULIS

Lynn Margulis was a major figure in the history of modern biology. Her theory of the origin of the eukaryotic cell is recognized as a great achievement. She predicted that the molecules would support her endosymbiotic theory. "Recent spectacular progress in deciphering the evolutionary significance of amino acid sequences in protein . . . suggests detailed homologies to the original eubacterium might even be discoverable" (Margulis 1970, 205). This suggestion and her prediction have been confirmed in a spectacular fashion. Not just amino acid sequences but the encoded messages of nucleic acid itself have been discovered. Both mitochondria and plastids contain DNA, and there is strong similarity (homology) between the DNA of these organelles and that of certain plausible bacteria of origin. The symbiotic origin of complex cells, or at least the bacterial origin of mitochondria and plastids, is now taken as a given. Margulis had a third proposal, namely, that the flagellae of certain cells—the whiplike appendages of some cells that function as motors for locomotion—likewise are the result of a symbiotic connection of more primitive cells. This idea has found less acceptance because flagella do not have DNA and do not share the microscopic structures of bacteria.

Margulis pursued symbiosis with enthusiasm—many would say beyond enthusiasm—for the forty years following her breakthrough with the evolution of eukaryotic cells. Because it is the paradigmatic case of union of some kind, she and her son became fascinated with the nature and causes of sex. She argued that sex in the higher organisms is, in a way, almost an afterthought. Endosymbiosis is but the tip of more general activity between bacteria as one unicellular organism tries to eat another. Sometimes the result is fusion without much ingestion, so there is a doubling of the gene content. Such larger cells can have a selective advantage, especially if conditions are difficult or harsh. But too much bulk, especially if there is repeated fusion, can be a bad thing, and so there are advantages to splitting. Something like this happening over and over again is the forerunner of sex. "Sex, like symbiosis, is a matter of merging. But it is also a matter of periodic escape from the merger. Sex can be understood as a very special case of cyclical symbiosis: both sex (fertilized eggs, the zygotes) and symbiosis, merging of symbiotic partners, produced new beings" (Margulis 1999, 130).

Few think that this picture is very plausible. Even fewer think plausible her related views about the origin of species (Margulis and Sagan 2002). A huge amount has been written on this topic, and in conventional evolutionary circles thought on the matter is still evolving. But, although there are various epicycles (for instance about the effects of small groups being isolated), the main deferent is Darwinian—groups of organisms get separated; they accumulate new mutations on which natural selection acts; and over time the groups evolve apart until, if brought back together, they can no longer interbreed. Obviously major empirical evidence of this would be that groups do evolve under the pressure of selection, and many studies confirm this. One of the most celebrated is that of the husband-and-wife team, Peter and Rosemary Grant, who have spent several decades looking at the finches in the Galapagos Archipelago (Grant and Grant 1989). Margulis was unimpressed: "They saw this big shift: the large-beaked birds going extinct, the small-beaked ones spreading all over the island and being selected for the kinds of seeds they eat. They saw lots of variation within a species, changes over time. But they never found any new species—ever." Indeed, on this matter Margulis, though not religious, was prepared to agree with the "intelligent design" critics of Darwinism, who claim that such "microevolution" never adds up to the "macroevolution" involved in the evolution of new groups. "The critics, including the creationist critics, are right about their criticism." Of course, Margulis disagreed with their religious conclusions, but she had her secular answer ready: symbiosis. "Numberless forms and variation come not just gradually and at random, but suddenly and forcefully, by the co-opting of strangers, the involvement and the folding of others—viral, bacterial, and eukaryotic—into ever more complex and miscegenous genomes. The acquisition of the reproducing other, of the microbe and its genome, is no mere sideshow. Attraction, merger, fusion, incorporation, cohabitation, recombination—both permanent and cyclical—and other forbidden couplings, are the main sources of Darwin's missing variation" (Margulis and Sagan 2002, 205). It isn't often that the person asked to write an introduction to a book states flatly that what follows is simply false, but Ernst Mayr, the distinguished evolutionist (who did more than any other to ferret out the details of speciation), was blunt: "There is no indication that any of the 10,000 species of birds or the 4,500 species of mammals originated by symbiosis" (xiii). In a like vein, the chair of the Department of Zoology at Oxford University, Paul Harvey (2004, 31), concluded after reading her book that he "could no longer take her seriously."

It was hardly chance that the one serious scientist willing from the first to align herself with Lovelock and the Gaia hypothesis was absolutely fanatical about symbiosis. Gaia sees the world as one, as an integrated organism. This was virtually a deduction from Margulis's world picture; she immediately supported the hypothesis and has stayed with it through all of the storms. She coauthored the initial papers with Lovelock and then continued to promote the idea. Writing with her son in 1984, she stated, "The Gaia hypothesis, presently a concern only for certain interdisciplinarians, may someday provide a new basis for a new ecology—and even become a household word. Already it is becoming the basis for a rich new world view" (Margulis and Sagan 1997, 146). One of her last arguments was that water is the key explanatory ingredient. "We champion the poorly developed Gaian view that life has vigorously helped maintain abundant water on the Earth's surface over the last three and a half thousand million years. We defend the idea that life's populations persist and continue to expand on Earth not because a 'lucky accident' has situated our moist planet at an optimal distance from the sun; rather communities of living organisms have actively maintained wet local surroundings. The result has been the retention of moist habitability over geological time" (Harding and Margulis 2009, 41). For instance, "exudates of microbial mat organisms directly retard evaporation," despite the fact that "elephant bodies carve out ponds and thus expose subsurface water to the surface" (52).

AUTOPOIESIS

Margulis thought little of advice that scientists ought to avoid philosophy, for it was in this realm that she tried to make an innovative contribution to thinking about Gaia. Somewhat independently but in some respects running parallel to Lovelock, she was concerned to see precisely what it might mean to think of something as an organism. In her inquiry, she encountered and strongly endorsed the work of Chilean thinkers Humberto R. Maturana and Francisco J. Varela. In order to characterize living beings, they have proposed the novel concept of "autopoiesis." This is not so much a turn from mechanism as a turn to organization. "*An autopoietic machine is a machine organized (defined as a unity) as a network of*

processes of production (transformation and destruction) of compo-
nents that produces the components which: (i) through their interac-
tions and transformations continuously regenerate and realize that
network of processes (relations) that produced them; and (ii) consti-
tute (the machine) as a concrete unity in the space in which they (the
components) exist by specifying the topological domain of its realiza-
tion as such a network" (Maturana and Varela 1980, 78–79, their
emphasis). It is rather difficult to make out what is at stake here,
but in a way the ultimate issue is not really whether it makes full
sense but what the authors (and anyone who supports them) take
it to mean. The basic idea seems to be that not only do organisms
regulate themselves but that somehow they maintain themselves
and regenerate parts that run down. They do this even though the
outside environment can change and impinge upon them. Also,
it is in this organization that we find the unity of the organism,
rather than something simply spatial. We have homeostasis in a
way, but it is more than merely mechanical. It is something that
captures the essence of being alive—organization that leads to a
new level of being and understanding. Emergence, in short.

One can see the attraction of this idea for a Gaia supporter of
Margulis's ilk. It is a—one might say *the*—crucial claim of the Gaia
hypothesis that not only does Earth maintain itself, but in some
way it can regenerate and respond in the face of change (e.g., the
increased heat from the sun). More than this, although obviously
Gaia is a spatial concept—it applies to Earth, for instance, and not
to the moon—it is in organization that we find the essence of Gaia.
Earth maintains its integrity as an integrated entity. It is homeostasis but a bit more. Margulis entered right into the spirit of all
this: "The smallest autopoietic systems, spherical [and] less than a
micrometre in diameter, are bacterial cells. (Viruses, plasmids, and
other replicons are too simple and small to be autopoietic.) The
largest autopoietic system, so far incapable of reproduction, is the
modulated surface of the Earth that Lovelock has named Gaia"
(Margulis and Sagan 1997, 98). And again, "The Gaian worldview
is an autopoietic one; the surface of this planet is alive with a con-

nected megametabolism, which leads to temperature and chemical modulation systems in which humans play a small and epiphenomenal part" (99).

Notice that Margulis put the emphasis on the living surface, not the whole Earth entity. This perhaps reflects a difference in interpretation between the very idea of a living being, an organism, and something more akin to a worldwide ecosystem.

> Lovelock would say that Earth is an organism. I disagree with this phraseology. No organism eats its own waste. I prefer to say that Earth is an ecosystem, one continuous enormous ecosystem composed of many component ecosystems. Lovelock's position is to let the people believe that Earth is an organism, because if they think it is just a pile of rocks they kick it, ignore it, and mistreat it. If they think Earth is an organism, they'll tend to treat it with respect. To me, this is a helpful cop-out, not science. Yet I do agree with Lovelock when he claims that most of the things scientists do are not science either. And I realize that by taking the stance he does he is more effective than I am in communicating Gaian ideas. (Brockman 1995)

Perhaps in this respect, therefore, the differences were not overwhelming. Yet it could be that in rejecting the organism talk, Margulis was dispelling concerns about teleology in order to open a path to what she did want to argue. For what made her position distinctive and surely separated her from Lovelock—showing that she was trying to go beyond mere homeostasis to a real organic unity—was where her thinking went from here, particularly with respect to Darwinian theory. We have seen that Lovelock got caught on the prongs of Darwinian selection. Dawkins and Doolittle both went after Gaia on the grounds that there is no reason to think that Earth (considered as a self-maintaining system) was produced by natural selection. Remember and reflect on Lovelock's response. In part, it was to deny that he was using the word *life* in the straight Darwinian sense of "produced by selection." He was using it more in the sense of a homeostatic system. In part, however, it was to accept the criticism and try to bring his thinking within the Darwinian boundary. This was the whole point of the Daisyworld model. Each flower was doing its thing without regard to or care for any other flower. Nevertheless, their collective actions led to stability. There was nothing there for Dawkins or Doolittle to object to. In other words, even if in some respects Gaia goes beyond Darwinism, it does not conflict with it. This at least was Lovelock's aim, even if Hamilton's reaction suggests that not everyone agreed that he was fully successful.

Margulis was not so accommodating. She saw Gaia, understood in an autopoietic sense, as being in conflict with Darwinism (meaning natural selection theory). She wrote of "big trouble in biology" and posited "physiological autopoiesis versus Darwinian mechanism." Although she was prepared to revere the name of Darwin himself, she was downright nasty about "the physics-centered philosophy of mechanism and its runt offspring neo-Darwinism" (Margulis and Sagan 1997, 271). (Neo-Darwinism is selection plus Mendelian or molecular genetics, and stresses the non-directionality of evolutionary change.) It is organization that counts, and it is organization that neo-Darwinism simply ignores. "The life-centered alternatives to mechanistic neo-Darwinism recognize that, of all the organisms on Earth today, only prokaryotes (bacteria) are individuals." Everything else demands group thinking. "All of the live beings ('organisms'—such as animals, plants, and fungi) are metabolically complex communities of the multitude of tightly organised beings. That is, what we generally accepted as an individual animal, such as a cow, is recognizable as a collection of various numbers and kinds of autopoietic entities that, functioning together, form the emergent entity—the cow. 'Individuals' are all diversities of co-evolving associates" (273).

Margulis would not bow to the individual-selectionist demands of Darwinism and wanted to stress the unity and cooperative nature of the evolutionary process. Selection must occur at all levels if (in line with some of the thinkers discussed in chapter 5) selection itself is always that crucial. Margulis also wanted in a non-Darwinian fashion to promote non-randomness, not in the sense of so-called intelligent design theory (supposing an outside intelligence) but of forward moves coming through integration and cooperation (as with the formation of the eukaryotic cell). Gaia is bound up with this: "Neo-Darwinists, who ignore chemical differences between living beings, who never factor autopoiesis into their equations, and who consider organisms as independent entities evolving by accumulation of chance mutations, must hate and resist autopoiesis and the Gaian worldview" (Margulis and Sagan 1997, 281). In short, the Gaia hypothesis and neo-Darwinism are rivals, not complements. This is fundamentally a philosophical disagreement. "If we can assume that consistency is a scientific virtue, then acceptance of a Gaia-autopoietic worldview requires that we reject the philosophical underpinnings of neo-Darwinism as it is currently practiced. Neo-Darwinism, in the Gaian perspective, must be intellectually dismissed as a minor, twentieth-century sect within the sprawling religious persuasion of Anglo-Saxon biology." There is a reductionistic, mechanistic, anti-holistic subtheme to Darwinism. For this reason, given also that this philosophy has vile social implications, it is to be rejected. "As yet another example of a thoughtstyle in the great family of biological-scientific *weltanschauungen*, past and present, neo-Darwinism (like phrenology and nineteenth-century German nature philosophy) must take its place (like British social Darwinism) as a quaint, but potentially dangerous, aberration" (281).

Margulis's enthusiasms, especially her support for the Gaia hypothesis—which Paul Harvey (2004, 31) referred to as "one of the weakest metaphors yet devised"—came at a cost. Unlike Lovelock, she was dependent on the world of academia, and so she suffered. In response to a 1992 grant application to NASA ("Microbial contributions to the pre Phanerozoic Earth"), the panel began with praise. "Margulis is a distinguished scientist who enjoys an international reputation. She is a prolific writer and a dynamic individual" (LC). It even went as far as to say, "She has been very successful in altering the way we think of life on Earth." But then came the stiletto: "It should also be noted that over the past several years she has gained a reputation, justified or not, of having gone perhaps too far. This is primarily due to her defense of the Gaia hypothesis." And so the funds and support remained meager to nonexistent. But Lynn Margulis was not deterred. From the days when she promoted in the face of huge opposition her beliefs about the origin of the eukaryotic cell to the moment of her death, she followed her convictions about the significance of unity, of parts coming together to make the whole. At an early age—because the boys seemed more interesting—she withdrew from her middle-class, private high school (the University of Chicago Laboratory School) and transferred to a significantly rougher public high school. She did not inform her parents of the shift. As far as she was concerned, decisions about what to do and what to think were going to be hers alone. That was always the case.

8

UNDERSTANDING

We draw to the end of our story. How are we to interpret it? My claim at the beginning was that, in order to make sense of the present, we had to understand the past. Let me now make good on this. We have two important scientists, James Lovelock and Lynn Margulis, a public background of people who were (and are) looking for an Earth-centered philosophy (perhaps even religion), and a professional scientific community (especially in biology) that reacted very negatively to the Gaia hypothesis. We must put everything in context to arrive at an overall picture.

JAMES LOVELOCK

We begin by thinking about Lovelock himself, the sort of man he was, the tradition from which he emerges, his background and achievements, and the moves he made through the Gaia controversy. Above all, remember that he comes from a very conventional scientific background in chemistry, with a practical bent—as is common in that science. He was educated in a no-nonsense way, and went on to do no-nonsense science, demonstrating outstanding talents in his field, particularly in the invention of important instruments for measuring various substances and their levels. Government and industry eagerly hired him because they needed his talents. His thinking and his skills were worth good money. The point is that Lovelock was rooted in the mechanistic, reductionistic tradition of Western science. He was taught to think and still thinks in a Cartesian fashion that matter as such is life-less. A molecule on its own is not a living entity; it is just piece of matter. Lovelock was taught and believed that matter is subject to unbending laws and that the secret to understanding it is in terms of proximate, efficient causes. There is no place for Aristotelian fi-nal causes in his chemistry. Above all, he thought in reductionistic terms. Small is beautiful. Remember that his greatest achievement, for which he is justly famous, was to invent a machine for mea-suring the presence of substances at unbelievably small levels—even to the domain of individual molecules. Lovelock was also a mechanist in the second sense, namely, that of thinking of the world as functioning in an machinelike way. From the first, he "wanted to play with machines," and his great talent as a scien-tist was to build models, gadgets. The Science Museum in South Kensington, his original place of inspiration, preserves many of his inventions (Hickman 2012). Even now, Lovelock still men-tions the 1960 Neville Shute novel, *The Trustee from the Toolroom*, about a man whose skill at designing and building models finds him friends all around the globe. "That's my world" (JL). When trying to understand physical reality, he projects just this kind of thinking. This is what you expect, he says, of "an old-fashioned scientist like me." I would amend this to say "old-fashioned, *Brit-ish* scientist like me."

Remember also that when Lovelock started to get interested in the special status of Earth, he did so in the context of finding that Earth is exceptional. Its atmospheric composition is simply not similar to that of other planets like Mars. He found this out by applying the laws of science, the laws of chemistry in particular, in innovative ways. He did not break with traditional science; he sim-ply did it better than anyone else. As he continued to study planet Earth, he persisted in applying only the laws of traditional physics and chemistry. In talking about the sulfur cycle, for instance, he used textbook science. The context may be innovative, but the chemistry is not. He would never propose a new and unorthodox chemical process. In fact, with regard to the whole of science, he is not into contradictions. If his thinking seems to violate the well-established ideas and theories of an area of science with which he is not familiar, then he worries. He is not Immanuel Velikovsky, the Russian-born psychoanalyst, who (in *Worlds in Collision*) argued that the planets had had different orbits and that their close encounters with Earth had caused all sorts of abnormal events (like the parting of the Red Sea), and whose causal explanations appealed to all kinds of unknown and unexpected electromagnetic forces. James Lovelock is basically a very conventional scientist. He was rightly elected to the Royal Society (of London)—a very conservative body with jealously protected standards for admission. All in all, I argue that Lovelock exemplifies those who accept and produce the mechanistic kind of science discussed in chapter 4.

Now, against this background—Lovelock the mechanist—put the Gaia hypothesis. It was conceived in the mid-sixties and reinforced when later he gazed on the photographs of our planet—they had a "spiritual effect" on him—and confirmed the insight that he was looking at a living being, at an organism. Certainly this was strange, but perhaps not extremely so. Although he was not formally religious, his own language reveals a spiritual side to Lovelock, probably going back to the childhood influence of Quakerism. "When I was a child I was marinated in Christian belief, and still it unconsciously guides my thinking and behaviour" (Lovelock 2009, 137). Quakerism puts a much greater emphasis on transcendental experience than on formal theology. Lovelock had broken with the Society of Friends not because of theological disillusionment—he looks back on Quakerism as being somewhat agnostic at best—but because his first wife loathed all religions and especially Quakerism for its pacifism—she disliked the "conchies" (conscientious objectors to war). Spiritual or not, Gaia is not a God substitute. "It sort of precludes religion almost. It's the atheist's dream in a way" (DS). But there is something there. "Science has taken over from religion as the authoritative source of information about life in the universe" (DS). Unfortunately, however, there is a cost: "Science has left a moral vacuum behind." And this is where Gaia has a role. "Gaia is important because it gave us something to which we were accountable." In short, "Because of that ethical significance, Gaia starts to become more than just a science. It begins to veer into that area previously occupied by religion" (DS). Julian Huxley wrote a book called *Religion without Revelation*. There are hints of such thinking here.

Although Lovelock's autobiography rather suggests that Gaia was born entire at one moment, the hypothesis took time to develop and mature. In an interview with Canadian scientist and journalist David Suzuki, Lovelock admits that the Gaia idea "took years and years to disentangle." Our ideas come by intuition. "You then have to spend years, first of all explaining to yourself what you've been thinking about. They're rather vague ideas at first." Then the real work begins. "Ideas began to come along about how this system could self-regulate itself. It took the best part of ten years" (DS). Lynn Margulis was important in the later part of this creative decade (they started working together in 1971); but in the earlier years, when Lovelock fully convinced himself that he was dealing with an Earth organism, he was encouraged and influenced primarily by the man who gave the hypothesis its name, the novelist William Golding.

In 1963, on returning from some years in America, the Lovelock family moved to the Wiltshire village of Bowerchalke, where they previously had had a weekend cottage. Although he had consultative work that still took him out into the wider world, Lovelock was now independent, no longer tied to an institute, university, or other organization that demanded daily group interactions. At home he had peace and quiet to think, work, and experiment. The village was chosen deliberately for its isolation, and the family lived there for ten years. During this time, Lovelock's one real friend— lots and lots of nights "chattering away" in the pub—was Golding, another rather isolated inhabitant of the village. "We were different. We sent our children to the village school"—as opposed to private schools, which is what middle-class English parents would have been inclined to do (BL). Lovelock was wrestling with the Gaia idea. "I was already beginning to look on the Earth as an organism. Or if not an organism, as a self-regulating system. Quite a complex one" (JL). And Golding was there to name, to clarify, to support with enthusiasm. "He was fascinated. That's why he suggested 'Gaia'" (JL). The idea was Lovelock's, but the facilitator was Golding. (Things might have turned out very differently had Lovelock's Bowerchalke neighbor been Richard Dawkins!) Golding suggested the name in 1967, so at least three or four years elapsed after Golding got interested before Lynn Margulis got involved.

There are independent reasons to expect that Golding would have been a sympathetic listener, encouraging Lovelock. Lovelock has suggested that it was because Golding knew some science. "He liked the theory, perhaps because he had been trained in physics as a student as well as in the arts" (OC). But there was more. The spiritual side of things was ever an obsession with Golding. This comes out in novel after novel, from the Augustinian themes of *Lord of the Flies* to the final trilogy, *To the Ends of the Earth*. Spiritual concerns were resonating with the times, especially in America, in parallel with the popularity of *Silent Spring*, and for many of the same reasons. Ten years after its original publication, and as the tragedy in Vietnam unfolded, *Lord of the Flies*, now a best-seller and essential reading in schools and universities, spoke to the evils of war and (inverting the imperialist themes of the nineteenth-century boys' yarn, *Coral Island*, that inspired it) of the contamination of unspoiled lands by the selfish culture of the West.

But there was a twist to Golding's spiritual concerns. He was no conventional Christian for, as a young man, he had been deeply influenced by the Rudolf Steiner movement, to the extent of living for a short while in a Steiner community house and teaching at a Waldorf school. He eventually broke from the movement, rejecting its more extreme views, but remained a close friend of one of the leading British members (Adam Bittlestone, who became a priest in the Steiner church, the Christian Community). In the words of Golding's recent, definitive biographer, anthroposophical ideas "left a permanent trace on Golding's beliefs" (Carey 2009, 49). More than a trace, actually. Take Golding's third novel, *Pincher Martin* (published in America as *The Two Deaths of Christopher*

Martin), written in the mid-1950s. It is a puzzling work. It tells of a naval officer in the Second World War, shipwrecked on a bare piece of rock, and of the subsequent week as he gets ever weaker, all the time thinking in flashbacks of his hurtful and immoral life, including an attempt to murder a rival in love (a character based loosely on Bittlestone). At the end of the novel, the reader learns that Martin died at once at the beginning of the week. What then of the flashbacks? The framework is a fictionalization of an essential piece of Steiner theology that concerns the *Doppelgänger*, our evil twin (Steiner 1914b). At the point of death, according to this idea, we encounter ourselves with all our faults. Martin, an actor before the war, remembers a play in which he has to "double" (Golding's language), playing the deadly sin of Greed. "He takes the best part, the best seat, the most money, the best notice, the best woman. He was born with his mouth and his flies open and both hands out to grab" (Golding 1956, 106). We need this recognition of our bad side, so that as we pass over into the spiritual world, we can recognize these faults, take account of them, and see things properly. The individual struggling through the week on the rock is the *Doppelgänger* breaking up, not the already-deceased Martin. "He continued to exist separately in a world composed of his own murderous nature" (Carey 2009, 196, quoting Golding in a letter to the *Radio Times*, March 21, 1958).

We need not assume that Golding took any of this literally to see that Steiner's metaphysics was an integral thread of his intellectual fabric. In reviewing a collection of anthroposophic essays, Golding praises Steiner for wanting to bridge the world of science and the world of spirit—"Most of us have an unexpressed faith that the bridge exists" (Carey 2009, 243). In another novel, *Free Fall* (1959), Golding says explicitly, "Both worlds are real. There is no bridge" (Carey 2009, 243). For one with Golding's background and interests, the idea of Earth as an organism was both familiar and comfortable. It was the very essence of the anthroposophical world view. Ehrenfried Pfeiffer, the doyen of biodynamic agriculturalists and the man who was so important for Rachel Carson: "In its totality, the earth is a living organism, the body of humanity, offering man a basis for his earthly development. It furnishes foodstuffs and raw materials, the stimuli for economic exchange. But this body of the earth offers still more; it offers a spiritual stimulus, it spurs humanity on to the completion of its tasks" (Pfeiffer 1947, 37). No wonder Golding responded positively when his friend Lovelock, a good scientist, proposed the idea independently. In Golding's acceptance speech for the Nobel Prize in Literature, Gaia figured prominently. We "have been caught up to see our earth, our mother, Gaia Mater, set like a jewel in space. We have no excuse now for supposing her riches inexhaustible nor the area we have to live on limitless because unbounded. We are the children of that great blue white jewel. Through our mother we are part of the solar system and part through that of the whole universe. In the blazing poetry of the fact we are children of the stars" (Golding 1983).

Lovelock, as we might expect, denies any knowledge of or direct influence by anthroposophy. As for Rachel Carson, there is the obvious motive for keeping any such connections quiet, and Lovelock would surely have felt uncomfortable with and scornful of much of the esoteric web of fantasy. Certainly, Lovelock lacked the empirical input that there was in the Rachel Carson case. But he had to know and approve something. At Golding's urging, for eight years starting around 1966, the Lovelocks sent one of their children (who was developmentally challenged) to a Waldorf school and thought very highly of it and its results. As we have seen, Steiner had a distinctive (to say the least) philosophy of education that permeated every moment of a Waldorf school day. It is very child-centered, thoroughly holistic, and (as Golding saw) entirely committed to bridging the world of spirit and of science. Art, based on Goethe's deeply non-reductionistic theory of colors, is very important; dance in the form of eurythmy, a kind of ethereal modern-dance form, stressing harmony and the "living spirit," is part of the school day; storytelling, often (as noted earlier) involving Germanic fairy stories incorporating large doses of magical happenings, is much encouraged; and science and nature study are, expectedly, inherently biodynamic. Given the effort that

Lovelock had to put into persuading his local educational authority to pay for his son's comparatively expensive education, it beggars belief that Lovelock knew nothing of or had no sympathy for the underlying metaphysics. He and his wife felt that there was something of value here—something in tune with his own world vision.

The Golding connection helps us to see why a scientist whose forte is building models and instruments should hold and articulate an idea so apparently alien to his background and talents. In one way, however, it was not that alien. He was thinking of the world in terms of a model. He was thinking of the world mechanically. He was thinking of the world in ways grounded in the scientific tradition from Hutton through Griggs and on to plate tectonics. In another way, perhaps, his vision was strange and alien, but thanks to his supportive friend, he found it exciting and compelling. If you ask how this paradox could be, it is worth noting that Lovelock was not steeped in the lore of Western civilization; indeed, he was rather ignorant of major intellectual traditions. I don't mean to imply that Lovelock didn't have a good science education, because he did. But he did not come from a background of wide culture. Plato was not a natural part of Lovelock's intellectual heritage, although he might have been had Lovelock gone to a private school (what the English misleadingly call a "public" school). He clearly did not know that in proposing the Gaia hypothesis, he was joining a tradition that came down from the Greeks and that had a vast literature. Amusingly and revealingly, when Golding first suggested "Gaia" as a name, Lovelock quite misheard him and thought he spoke of something else. "I have no classics" (OC). He was mostly unaware also of the battles that had taken place during the Scientific Revolution over issues to do with final causes, or of the ways in which Darwin did or did not expel them from biology. In matters like this, he clearly had no idea of what he was embarking upon. He certainly never thought that anyone would support his ideas because of Gaia's religious virtues. Again, the mechanistic side to things comes forward: "Here you've got this thing running itself. It doesn't need God interfering" (JL).

Combine all of this with a kind of public, cocky self-assurance: hubris. Lovelock makes much of the fact that he stands outside the usual academic life-path; that he has made his living by his wits and his inventions; that he distrusts granting agencies with their anonymous refereeing; and that in response to harsh judgments he has shown with great satisfaction that he alone truly was right. As it often is, this attitude is combined with a deep, private sense of insecurity, of low self-worth. According to his first wife, Jim "had . . . a poor self-image" (Lovelock 2009). He mixes with people better educated than he is and who have smoother social skills. He admits that at least part of the reason he stands outside the academic system is that he is not sure he would be respected enough to succeed. He is not arrogant but more "bolshie." Hence he would set himself to defy the establishment, the too-confident Richard Dawkins types from Oxbridge, and beat them at their own game. On the one hand, this helped him in the Gaia controversy. He was independent. He had been proven right in the past when everyone said he was wrong. If people sneered at him, it was to be expected that they would treat an outsider such as he in this way. That is why the friendship and support of Golding, a man who seems also to have been somewhat at odds with his society, was so important. But on the other hand, his stance did contribute to his isolation, especially in an intellectual sense. Because of his ignorance, and perhaps because of his insensitivity to the battles of the past and of the way that culture, including science, had matured, he was (especially in the 1970s) rather given to statements that he later regretted. I mean especially how he slipped right into the language of ends and purposes as he promoted his vision of Earth as a living being. He was a mechanist, but he didn't always talk like one. In his lack of sophistication about the ins and outs of final-cause thinking, he did not always even think like one. Paradoxically, Lovelock's particular relationship to mechanism may have been important here. Remember that Lovelock is not a deep thinker, in the sense of conceiving of and developing fundamental theories of science. He is a tinkerer, an inventor, an instrument maker—and such people do think teleologically. You make the electron capture detector in order to detect very low traces of chemicals. It is true that in thinking about models in science—mechanism in the second sense—you take the teleology out. Ever-circulating continental plates have no purpose, no ends. But if your model-building is not to become part of the theory of science, ends taken off or out, but to lead to instruments for which people will pay hard cash, it would not be surprising that when you do lift your eyes from the workbench to look at the stars—or at pictures of our planet from space—you are more committed to purposes than most of your fellow mechanists would find acceptable. There is a whiff of Platonism there, after all.

Recall what did happen when Lovelock was criticized by the likes of Dawkins and Doolittle, the Darwinians. It is true that he turned on his tormentors, saying that they did not appreciate the complexity of the case and that, as conventional scholars, they dared not break out from conformity. What he did not do was attack mechanistic, reductionist science. Let me explain what this means, because in some respects you might think that this is precisely what Lovelock did. Surely his major move was to cross the divide to holism, to emergentism, to the science of chapter 5, the science of the organismic biologist. Remember, he argued that a perfectly respectable conception of life need not refer directly to natural selection. It can be framed in terms of homeostasis. And this, as we have seen, is the central notion of the unambiguously holistic scientist W. B. Cannon. Lovelock adopted those terms and emphasized them. For instance, in a commentary in *Nature*, under the heading "Holism," he wrote that the Gaia hypothesis says that the "close coupling of organisms and their environment is strong enough to have greatly influenced the way in which the life-environment system on earth, and on other planets with life, has evolved. It is strong enough that we will not properly understand Earth history until we think of the system as just that, our whole system, and stop trying to understand its parts in isolation from one another" (Lovelock 1990, 101). Elsewhere he wrote, "In many ways Gaia, like an invention, is difficult to describe. The nearest I can reach is to say that Gaia is an evolving system, a system made up from all living things and their surface environment, the oceans, atmosphere, and crustal rocks, the two parts tightly coupled and indivisible. It is an 'emergent domain'—a system that has emerged from the reciprocal evolution of organisms and their environment over the eons of life on Earth" (Lovelock 1991, 11).

Nevertheless, we should take these comments with a strong dose of caution. As we know, there are different meanings, different degrees, to holism and emergentism. Certainly, in the weakest sense, these are not merely compatible with but practically necessitated by mechanism. A clock is holistic, an emergent system in this sense, and the clock is the paradigm for the mechanist. The stronger notions of holism and emergentism go from the unexpected (who would have thought that two such dangerous substances as sodium and chlorine could make inoffensive table salt?) to the nearly impossible (how can something made of meat give rise to consciousness, to sentience?). Whatever language he may be using, and notwithstanding that at some level Lovelock is using Gaia to give a metaphysical world picture, he gives no reason to think that he breaks from mechanism at all in these tougher, more demanding respects. His notion of an emergent system is a feedback loop. Completely mechanistic. And his notion of holism hardly goes beyond an exhortation to consider all of the facts and think them as integrated, as "not being limited by a particular science" (JL). You cannot investigate the temperature of the ocean without taking into account the sun, the laws of evaporation, the growth of algae, and so forth. Admittedly this is all tremendously important, but it is no more than an explicit mechanist like Richard Dawkins would insist upon.

Further support for the point being argued, that essentially James Lovelock is a mechanist, is that when he was challenged, he came up with Daisyworld, an ultra-mechanistic, causal explanation as to how homeostasis can be achieved. Incidentally, it is about as machinelike (that is, mechanistic in the second sense) as one could imagine. It is in line with other feedback machines, such as steam engines with regulating governors and electric irons, to use two examples that were mentioned. And not so incidentally, it also lent itself to computer modeling. Daisyworld is a conceptual machine that thrives on physical machines—Lovelock's kind of science. Moreover, in its basic details, Daisyworld is as thoroughly grounded in individual selection as one can imagine. It involves no nonsense about things happening for the good of the group. It really does pass the Richard Dawkins test of acceptability. No wonder that William Hamilton felt able to appreciate it in certain respects—especially when Lovelock agreed with Hamilton's rejection of putative alternatives to selection as sources of adaptation. Hamilton had written that "today complexity theory is often invoked as a route to adaptation that is not based upon natural selection. It has, however, yet to show how any adaptation can arise at all" (Hamilton 1998). Lovelock responded, "I also agree that complexity theory confuses rather than helps our understanding of the link between Gaia and natural selection" (Lovelock 1998).

Remember also that by this time Lovelock was repeatedly stressing that he was using a metaphor here, as others did also, and that doing so is perfectly acceptable for a reputable scientist. He wrote, "I often describe the planetary ecosystem, Gaia, as alive, because it behaves like a living organism to the extent that temperature and chemical composition are actively kept constant in the face of perturbations. When I do, I am well aware that the term itself is metaphorical and that the Earth is not alive in the same way as you or me, or even a bacterium" (Lovelock 1991, 6). I do not mean to say that this satisfied the critics. It did not. Hamilton argued that feedback loops as such did not necessarily produce the kind of homeostasis being posited. As pointed out in chapter 4, for the Darwinian, some kind of natural final cause can only arise when selection has produced it—that is why Hamilton, like Dawkins, kept stressing that Gaia failed because there was no selection at the planetary level. However, the point here is that Lovelock thought he had done the job. Not being a biologist, he was simply not sensitive to the consequence spotted by Darwin and endorsed by Hamilton, namely, that in a world of individual-based natural selection, notions like the balance of nature, homeostasis (especially as applied to larger systems), and equilibrium are under threat and can never be the fundamental forces of change. The door to that world has been shut.

Finally, let us mention a major piece of history that Lovelock absorbed and to which he makes constant, self-supporting reference: the characterization by the eighteenth-century Scottish geologist James Hutton of Earth as an organism. In *Theory of the Earth* Hutton speaks of Earth as an "organized body," and introduces this notion to explain how the planet persists (indefinitely, as far as Hutton is concerned). Earth "has a constitution in which the necessary decay of the machine is naturally repaired, in the exertion of those productive powers by which it had been formed" (Hutton 1795, chap. 1, sec. 1). But this is only part of the story. In the middle of this quoted passage, we have the weasel word *machine*. As we saw in chapter 4, the overriding hypothesis endorsed by Hutton is that the world is a machine devised and created by God. "When we trace the parts of which this terrestrial system is composed, and when we view the general connection of those several parts, the whole presents a machine of a peculiar construction by which it is adapted to a certain end. We perceive a fabric, erected in wisdom, to obtain a purpose worthy of the power that is apparent in the production of it" (chap. 1, sec. 1). For James Hutton, the world is a Newcomen engine, and that is pretty much the stance—brought up to date by modern technology—of James Lovelock. He is firmly in the Western mechanistic, reductionistic tradition. He explains things in terms of constant laws; he thinks in terms of models, machines, although this does sometimes backfire.

LYNN MARGULIS

Although Lynn Margulis certainly supported Lovelock, her background and stance are very different from his. She truly does belong with the people of chapter 5, the holists, the emergentists, the organicists. Yet she was not in any sense religious. Initially, she opposed the name *Gaia* on grounds that it might be taken this way. She was more astute than Lovelock on this matter. "It kind of brought up the notion of Pagan goddesses and things, which didn't fit at all with her atheist view of science" (DS). But her philosophical views were quite different from Lovelock's. This explains why, after those first collaborative papers, Lovelock and Margulis never wrote together again. Although Lovelock refers repeatedly to the unfulfilled nature of his first marriage, there was not (as one might suspect) an emotional entanglement and subsequent breakup. They really were just good friends and, until Margulis's death, they were incredibly supportive of each other. Rather, they were scientists of very different schools. "I drifted apart from Lynn—no fight or breakup or bad feelings—it was just that I could not get Lynn to understand Gaia, the theory of it. She had no trust whatever in models or mathematics; she thought it was just useless. That's not a good way to collaborate with a physical chemist" (JL).

Even before she got to college, when she was at the University of Chicago Laboratory School, Lynn Margulis was being fed the holistic message. Judging by the texts written for children—for instance, *Every Day Is Earth Day*—one of her most admired and inspiring teachers was the head of the science unit, Illa Podendorf. She thought entirely in a kind of proto-Gaia fashion, as in this passage from a text she wrote for her students: "Sometimes people dump things into a lake that make changes in it. Then strange plants grow and thrive in the water. A lake is no longer balanced when it is choked with plants. The fish must swim to another place or die" (Podendorf 1971, 22). After this, the integrative perspective of the Great Books Program at the University of Chicago in the 1950s had a deep and lasting influence. The pioneering president Robert Hutchins had resigned in 1950, but his neo-Aristotelian influence lived on. First, the faculty encouraged a willingness to think outside the box. "Science at the University of Chicago was superb—no set methods, honest, open, and energetic, that facilitated asking the really interesting questions of philosophy. 'What are we?' 'Where do we come from?' 'How do we work?' 'What is this universe?' I never for a moment doubt that I owe my choice of a career in science to the wisdom of the Chicago education" (LC).

Second, there was an interest in conjunction, symbiosis, holism. Chicago's biology department was in transition. Allee had retired in 1950, and Sewall Wright would retire in 1955. But although molecular biology was on the horizon, it did not become established at the university until the 1960s, when the molecular biologist George Beadle was appointed president (Sloan, unpublished paper, 2012). Back then, students did not choose a departmentally affiliated undergraduate major, but the philosophy of Chicago's older generation biologists was still the pedagogical foundation. Texts were such tomes as Ralph Buchsbaum's *Animals without Backbones*, fulsome in its thanks to and praise of Allee and Emerson (JC). "The theme of my 'Nat. Sci. 2' class was 'What is heredity? What links the generations? How do the materials in fused egg and sperm inspire the development of an entire animal? How does variation in evolution arise?' These haunting questions drive me still" (Margulis 1999, 23). And always there was the exhortation to think of science in terms of underlying themes, patterns, and approaches. "We were taught how, through science, we could go about answering important philosophical questions" (23). The students worked on Mendel's peas, trying to discover how the different colored peas pass on their traits from one set of plants to the next, sometimes concealing colors for one or more generations. They worked on stentors, microscopic unicellular organisms that can regenerate (self-organize) when divided and can live symbiotically with algae. Particularly significant for Margulis was Vance Tartar's work that involved grafting one form of these organisms onto another, trying to show non-Mendelian inheritance. "Even as an undergraduate, I sensed something was too pat, too reductionistic, too limiting about the idea that genes in the nucleus determine all the characteristics of plants and animals" (24). Such thinking always strives to get away from the blind, mechanistic purposelessness of Darwinism. "How could random gene mutations lead to the evolution of flowers and eyes?" (LC) For Margulis, selection fueled by harsh competition was not the answer. One of her instructors was Beatrice Mintz, who was soon to make path-breaking discoveries involving transgenic mice (UC). Margulis noted, "I've been working ever since on the idea that variation in evolution arises by symbiosis" (LC).

Lynn Margulis falls firmly within the transplanted-to-America, holistic-emergent tradition in a full and (obviously) very fruitful way. Her successful theory of the origin of the eukaryotic cell depended on a number of factors. Of vital importance was the time she spent after Chicago in Wisconsin, where she completed a master's degree before moving on to doctoral work at Berkeley. She was thoroughly exposed to work on heredity taking place outside the nucleus (Keller 1986). One of her professors at Wisconsin, Hans Ris, was to make the discovery of DNA in the non-nuclear chloroplasts, and, sensing that this might be deeply significant, he even went so far as to read aloud to a group of students (including Margulis) earlier-in-the-century speculations that complex cell parts might have originated as freestanding simple cells (Margulis 2005, 146). Margulis was thrilled at the idea, and, in her own words, "the course of my professional life was set forever!" What distinguished her from the other listeners was that Ris's intuitions fell on fertile philosophical ground. She incorporated into her theory thoughts of union, of symbiosis, between two or more different prokaryotic cells. And that was clearly the path she followed in the four ensuing decades. Gaia, flagella, species, sex. Things coming together and creating something new. It is not that mechanistic science was absent from Margulis's world picture. She wanted to follow the laws of chemistry no less than does Lovelock. Nor did she want to ride roughshod over every theory outside her immediate orbit. She was no "young Earth creationist," for example. She was as committed to evolution as was Charles Darwin. But there was an altogether different flavor to her work, and (as we have seen) if a theory like Darwinian evolution through natural selection stood in her way, her strategy (unlike that of Lovelock) was not to mold and compromise, but to push and reject. Her way of doing science was not that of her co-enthusiast for Gaia. Nor was it that of G. Evelyn Hutchinson and his school, even though he had written a supportive foreword to her book on the origin of eukaryotic cells, a favor that surely led to the favorable references to his work in the early Gaia papers (Hutchinson 1970).

Does this add up to anything significant? She certainly thought that it did, and she was certainly genuine in this and not half–torn, as one senses in Lovelock at times. It shows above all in her approach to scientific issues. Here she was fundamentally and thoroughly holistic. She was always looking for connections, for the way that things link up rather than break down. "How do the materials in fused egg and sperm inspire the development of an entire animal?" (LC) She thought that this gave meaning in a way that a reductionistic world picture did not. A selfish-gene theory of evolution had to be at best incomplete, and at worst fundamentally misleading and wrong. Her holism showed in the kind of understanding she had of science, especially of the science she had produced or endorsed. This point is intimately tied up with her enthusiasm for the autopoiesis hypothesis—one that many commentators note owes nothing to Anglo-Saxon thought and everything to German idealism. Maturana and Varela do not want to deny mechanism as such. They speak in terms of "living machines," and we have seen that the notion of autopoiesis is introduced in the machine context, as in "an autopoietic machine is a machine organized (defined as a unity)." They also want to avoid vitalism and final causes. "Living systems, as physical autopoietic machines, are purposeless systems" (Maturana and Varela 1980, 86). However, organization does seem to imply some kind of new dimension of understanding. The language is not clear, but this is surely the intent of the following: "A phenomenological domain is defined by the properties of the unity or unities that constitute it, either singly or collectively through their transformations or interactions. Thus, whenever a unity is defined, or a class or classes of unities are established which can undergo transformations or interactions, a phenomenological domain is defined" (116).

Certainly, it is in this non-reductionistic, emergentist mode that Margulis located herself. She went so far as to say of autopoietic systems that "unlike mechanistic systems, they produce and maintain their own boundaries" (Margulis and Sagan 1997, 348)—self-organization! And, arguing against a specific call for reductionism, she wrote, "We compare this pervasive mechanistic belief of biologists, most of whom are smitten by physicomathematics envy, with a life-centered alternative worldview called autopoiesis, which rejects the concept of a mechanical universe known by an objective observer" (266). She wanted to go further down the path than Maturana and Varela, arguing not that her holism was to be added to the top of a mechanical world picture, but that it challenges and refutes such a view. And it is clear that her understanding of Gaia—where she would locate Gaian understanding—was outside the mechanical orbit. For this reason, she did not hesitate to breach the is/ought barrier so beloved of thinkers in the mechanistic tradition—the barrier (spelled out by David Hume) that says you cannot go from claims about matters of fact to claims about matters of obligation. Morality is not just science writ large. Margulis brushed this aside, showing herself to be rooted in the tradition of all of those neo-Spencerian, American holists (down to and including Edward O. Wilson) who happily thought of human society in terms of their biology: "There is something fresh, new, and yet methodologically appealing about Gaia, however. A scientific theory of an earth that in some sense feels and responds is welcome." Elaborating on this, she wrote, "The Gaian blending of organisms and environment into one, wherein the atmosphere is an extension of the biosphere, is a modern rationalist formulation of an ancient intuitive sentiment. One implication is that there may be a strong biogeological precedent for the time-honored political and mystical goal of peaceful coexistence and world unity" (156). What would you expect of a woman whose undergraduate education included "lots of Plato" (LM)?

The Russian scientist Vladimir I. Vernadsky (1863–1945) promoted the idea of the "biosphere," in which life (the "noösphere") influenced the course of geology. (Pierre Teilhard de Chardin, the French Jesuit paleontologist, reconciler of science and religion, picked up the notion of the noösphere from attending Vernadsky's lectures, when the latter was living in Paris.) When formulating the Gaia hypothesis, neither Lovelock nor Margulis had heard of Vernadsky and his work. When they learned of it, both were naturally delighted to find another predecessor, although Lovelock is more reserved about the real significance of the biosphere ideas. Margulis, however, was enthusiastic, writing (with others) the introduction to the full translation of *The Biosphere*. This made good sense because, like a good organicist, Vernadsky took an explicitly holistic position between mechanism and vitalism. Mechanism failed to capture the "complexity of phenomena." Thus, "the living organism of the biosphere should now be studied empirically, as a particular body that cannot be entirely reduced to known physicochemical systems" (Vernadsky 1998, 52). Francisco Varela, naturally, wrote a blurb for the book.

Let us put Margulis into context. She was as authentic as Lovelock. He stands in the mechanist tradition. She stood in the organicist/holist/emergentist tradition. They may have been partners in developing Gaia, but they were far from partners in their philosophies of science. She was part of that anti-reductionistic, mechanism-wary tradition that goes back into the nineteenth century and before—a tradition that was a significant part of American biological life in the twentieth century (remember Gould and Wilson). And bear in mind what has been stressed repeatedly in this discussion: that the side one takes influences the science one does and the science one finds acceptable. The holist/emergentist is going to be looking for groups, for wholes, in a way that the mechanist/reductionist is not. And the kinds of theories acceptable to the holist/emergentist are going to be different from those acceptable to the mechanist/reductionist. We have seen this point illustrated starkly in the different attitudes taken toward the individualistic-competitive nature of Darwinism: the mechanist/reductionists endorse it, and the holist/emergentists reject it. In the immortal words of Private Willis in Gilbert and Sullivan's *Iolanthe*:

I often think it's comical—Fal, lal, la!
How Nature always does contrive—Fal, lal, la!
That every boy and every gal

That's born into the world alive
Is either a little Liberal
Or else a little Conservative!
Fal, lal, la!

You can say much the same about mechanists and organicists when it comes to individual selection versus group selection. Their allegiance is beyond reason and evidence. It is all about commitments, philosophies. When we consider Margulis's position in the light of this, we see how unambiguously and naturally she falls into the holistic/emergentist tradition. She aligns herself with those for whom integrative processes are significant in their own right. And if there is teleology in her thinking, as I have earlier suggested may be true of holists, then it is (as opposed to Lovelock's teleology) the Aristotelian kind of thinking about final causes. Fal, lal, la!

THE WORLD AT LARGE

The enthusiastic reception of the Gaia hypothesis by the general public, especially the general American public, was not surprising. Hylozoism, the belief that Earth is an organism, is a tradition of thinking about the world and its environment that was especially strong in America, going back to the transcendentalists. It is a fundamental part of that philosophy. When Rachel Carson, who ought to have supported it, did not do so, we suspect that she had some reason to conceal what apparently was a firm conviction, and, as we saw, the suspicion proves correct. She needed to conceal some of her sources to avoid being judged a kook, a purveyor of mere pseudoscience.

Grant, then, that the public was delighted. But who or what was this public? Was it really "general"? We can surely assume that it included the more left-wing, socially progressive people in America and also in other countries like Canada and the United Kingdom. Among the young, especially, this was a popular idea. But—and again one thinks of America in particular—what about the churchgoing public, especially the Protestant churchgoing public? Lynn White (1967) hit a nerve. There were and are good reasons why many would reject or at least suspect the Gaia hypothesis on theological grounds—especially if there is any hint that we have moral duties toward Earth in itself. According to the evangelical Christian head of biology at Wheaton College, Billy Graham's alma mater, "Scripture provides a logical value system. It establishes that the whole creation in general, and every part of it in particular, has a value given to it by God. This does not mean that the creation is inherently good or that it has the right to exist on its own merits, independent of God. Its goodness is derived from its Creator and so is a kind of "grace" goodness, freely given in love, not grudgingly merited by right" (Van Dyke et al. 1996, 53). Likewise, the Lutherans say, "The earth is very good. Neither demonic nor divine, neither meaningless nor sufficient unto itself, it receives its meaning and value from God" (Evangelical Lutheran Church in America 1996, 245). This means that one who wants to avoid offending certain religious groups must be careful about notions like the Gaia hypothesis. "Though the hypothesis itself can be considered reasonably scientific, it has spawned a host of ideas and philosophies which reach out to deify the earth" (Van Dyke et al. 1996, 139). This is idolatry. It would be unscriptural to worship the earth. "Creation worships the Creator" (ELCA 1996, 244).

There were some who wrestled with these issues, trying to see a place in the Christian schema not just for environmentalism but for Earth considered as a living entity in its own right. Such efforts tended to come from those who were not so dogmatically committed to the absolute sovereignty of God and who did not think that *sola scriptura* begins and ends the discussion. Thomas Berry, a Catholic priest, made the organic thesis central to his theological vision. "The universe is not a vast smudge of matter, some jelly-like substance extended indefinitely in space. Nor is the universe a collection of unrelated particles. The universe is, rather, a vast multiplicity of individual realities with both qualitative and quantitative differences, all in the spiritual-physical community with one another" (Berry 2009, 71–72). Such integration ties in explicitly with the idea of Earth as an organism. "This unique mode of Earth-being is expressed primarily in the number and diversity of living forms that exist on Earth, living forms so integral to one another and with the structure and functioning of the planet that we can appropriately speak of Earth as a 'Living Planet'" (110). Making the case complete, Berry included familiar criticism of Darwin for his overemphasis of struggle and survival. "Darwin had only a minimal awareness of the cooperative and mutual dependence of each form of life on the other forms of life. This is remarkable: he himself discovered the great web of life, yet he did not have a full appreciation of the principle of intercommunion" (73). Similar sympathy for a living earth exists elsewhere among the faithful, including (perhaps surprisingly) the theology of the Church of the Latter Day Saints of Jesus Christ. We learn from *The Book of Moses* (dictated to Joseph Smith in 1830 and 1831, and now incorporated in *The Pearl of Great Price*, one of the four sacred books of the Mormon canon) that Earth itself is quite able to express fairly strong emotions: "Wo, wo is me, the mother of men; I am pained, I am weary, because of the wickedness of my children. When shall I crest, and be cleansed from the filthiness which is gone forth out of me? When will my Creator sanctify me, that I may rest, and righteousness for a season abide upon my face?" (7.48).

Of course, some may feel that I am selecting passages that support my thesis. Catholics have long had social activists within their ranks, and it is no surprise that this would extend to matters environmental. The Mormons, however, tend to be socially, philosophically, and politically conservative and have not been great friends of environmentalism. The point is not that everyone in the general public would have welcomed Gaia, but that we should not assume that only people of the more extreme left or radical movements would welcome it. Yet the Gaia hypothesis obviously did and still does on balance appeal more to those challenging conventional norms. We can begin more or less in the center with people like Mary Midgley, who are fully professional and established in their disciplines (in her instance, philosophy), but who have strong organicist leanings. Whether she thinks of the hypothesis as merely metaphorical (recognizing that good metaphors are rarely "mere") or as something more literal, it is easy to see why she embraced the hypothesis, finding it almost a natural deductive consequence of her world picture. Then we move out to the radical groups like the deep ecologists and somewhat related groups like the ecofeminists. Finally, on the fringe we find anthroposophists and their sympathizers at one end, prepared to suggest that Rudolf Steiner was the reincarnation of Plato. Charles, Prince of Wales, who talks to plants, comes in the middle, happily quoting the alchemic "Emerald Tablet of Hermes": "And as all things are One, so all things have their birth from this One Thing by adaptation. Its power is integrating if it be turned into Earth" (Prince of Wales, Juniper, and Skelly 2010, 120). Pagans and fellow travelers are found at the other end. "You and I and every other person, creature, tree, and flower are cells in the greater living body of Mother Earth—or Gaea, as many call Her. And the living Earth is only one of countless bodies of all sizes—planets, moons, asteroids, comets, meteorites, planetoids, and planetesimals—that make up our solar system" (Zell-Ravenheart 2004, 53). We are a tiny part of a tiny part, but despite our insignificance, ultimately all is made whole and meaningful. We end as we began, back with Plato: "All of it—every atom in your body; you and all your family, friends, and neighbors throughout the world; every living creature and plant upon the face of the earth; every planet, moon, and comet in the solar system; every star in the Milky Way galaxy, every galaxy in the vast and infinite universe—all are connected into the one great Web of Unity, one great Universal 'Internet' of Space and Time, Matter, and Energy" (52).

THE PROFESSIONAL SCIENTISTS

That Lovelock was surprised by the public's embrace of Gaia only goes to show what I have emphasized earlier—he simply wasn't in the same tradition as people like Midgley, let alone the ecofeminists and Pagans. Margulis had a much better sense of the conceptual geography of the times. Nevertheless, Lovelock (more than

Margulis) was ready to befriend folk across a broad spectrum, de-lighted at almost any support. If churches wanted to have a ser-mon from him, he was game. If Zell-Ravenheart wanted to enter into correspondence, he was glad to oblige. If someone wanted a foreword to a book, look no further. In a way, though, Love-lock could be so accommodating because the public realm was not really where his heart lay. Like Charles Darwin a hundred years before him, he appreciated public adulation, but it was the profes-sional community that counted. A remark about New Age think-ers reveals that his friendliness is more a facet of his warm person-ality than of intellectual conviction: "I get very cross. No time for them at all" (JL). As intimated above, many New Age thinkers were using Lovelock for their own ends rather than appreciating him in his own right. They were more truly Platonic in seeing the whole universe throbbing with life, whereas for Lovelock, the key is that Earth is unique. (Interestingly, Lovelock apparently has suf-ficient knowledge of anthroposophy that [probably accurately] he does not think Rudolf Steiner's Platonic inclinations truly align him with the New Age movement.)

We move now toward the final part of our story. Why did the professional scientific community turn so violently against one of their own? Why didn't Lovelock and Margulis get a full and re-spectful hearing, with a genuine attempt to find the gold in the dross? Part of the answer is that they were heard, that they did get a full examination, and that in the opinion of many they failed the test. Gaia didn't do what professional scientists expect of a good, new, fruitful hypothesis like evolution through natural selection or continental drift through plate tectonics. It didn't lead to new pre-dictions, ideas, connections, and so forth. As Kirchner said, "Gaia may be a grand vision, but it is not the kind of vision that can be scientifically validated" (Kirchner 1989, 233–34). More recently, we hear, "To be kind, it could be said the Gaia hypothesis is a sort of metaphor. More to the point, it is pseudo-science devoid of explanatory power" (Fenchel 2003, 149). Obviously, Lovelock would have challenged this. He felt with some reason that he was making a successful effort to show that Gaia could lead to new and exciting work. The paper on sulfur in a feedback loop controlling Earth's temperature is a case in point. Whether or not this and related work managed to justify the claims made for Gaia is another matter. But even if it was not enough, this was hardly reason in itself for nastiness.

Stephen Jay Gould, like the critic just quoted, complained that Gaia at best is merely metaphorical. But even if this is true, it is hardly cause for rejection and scorn, especially not from Gould, a master of scientific metaphor (e.g., punctuated equilibrium, spandrels, *Baupläne*). The passage quoted earlier from Gould about metaphor, now given in its entirety, is highly revealing: "I am especially wary of arguments that find kindness, mutuality, synergism, harmony—the very elements we strive mightily, and so often unsuccessfully, to put into our own lives—intrinsically in nature. I see no evidence for Teilhard's noösphere [something praised by Tim Zell], for Capra's California style of holism [Capra is a deep ecologist], for Sheldrake's morphic resonance [which led the editor of *Nature* to call for book burning]. Gaia strikes me as a metaphor, not a mechanism" (Gould 1987, 21). What Gould is picking out and linking to Gaia (as did the critic just quoted) is pseudoscience. Remember, this was the label attached to Sheldrake. As a professional scientist, Gould detested those he thought were charlatans. He wanted nothing to do with systems whose justification is only an ideology, a wish to find in nature precisely what can be used to buttress one's own convictions, which have been read into nature in the first place. He felt this way especially about systems that included any hint of teleology.

By the end of the 1970s, Gaia was no longer a rather mild viewpoint, beloved of environmentalists and others in the post-transcendentalist ecological tradition. Everybody knew about the California flower children, with their chants, herbs, and various hallucinogenic substances. Everybody knew about alternative medicines, philosophies, and religions—deep ecology, ecofeminism, Native American spirituality. Everybody knew about various nutty practices and ceremonies. Purify yourself with bathing and clean clothes, go through the right rituals and incantations, and feel yourself at one with the universe. Everybody knew about the weird cults, be they Scientologists, or Hare Krishna, or Pagans. And everybody knew (or assumed) that there was sex, sex, sex. Gaia was embraced enthusiastically by this culture—something that Lovelock too often gave the impression of relishing. No wonder the professional scientists were horrified. It is often not the content but the company it keeps that renders a concept pseudoscientific. Gaia was being taken up by a man who called himself "Oberon" (who had previously been "Otter"), who at the time was in the unicorn breeding business (actually goats with surgically altered horns), was about to set off to the South Seas in search of mermaids, was living in a sexually liberated ménage that went beyond every adolescent boy's fantasies, was practicing a religion that liked nothing so much as for its practitioners to go around in their birthday suits (a room full of naked Pagans working the ditto machines to produce *Green Egg* is indeed a sight for sore eyes), and who was into "drawing down the Moon": "Speak to the Goddess that is the Moon, asking her to fill you. Feel the energy as it flows down your arms and fills your soul" (Zell-Ravenheart 2004, 200). It is no wonder that the scientific community reacted as though a bad smell had been let off at the vicar's tea party.

You might think that this is unfortunate and that I exaggerate. After all, scientists generally tend to be people of the left (a 2009 survey in the United States found only 6% of them registered Republican as opposed to 87% Democrat or Independent)—and supportive of environmental movements and the like—84% of scientists and only 49% of the general public think that humans contribute to global warming (Mervis 2009, 132). Surely scientists should have seen beyond the culture of the time and appreciated the worth of the Gaia hypothesis. But scientists, although they may have certain liberal tendencies, are in other important respects very conservative people—a bit like Episcopalians. They have been selected for their talents. They have had long and often arduous training. They have a place in society that depends on their skills and on their not breaking rank. They have an ideology about their practices and integrity. (Note the fondness of scientists for the philosophy of Karl Popper, which privileges and honors the fearless scientists who ruthlessly put their hypotheses to the test and reject those that are inadequate.) They define themselves in major ways by refusing to be pseudoscientists. They resent people who try to achieve the status of "scientist" without the work, talent, and persistence.

PROFESSIONAL SCIENCE UNDER THREAT

There is a final piece to the story. Assigning something to the category of pseudoscience does not mean that it must remain there permanently. Although epistemological factors are important (How consistent are the claims with those of other more accepted sciences? How good are the predictions made on the basis of the theory?), sociology and psychology enter in as well. Especially important is the extent to which professional scientists are feeling tense or under threat from other causes (Gordin 2012; Ruse 2013b). At times when they fear that their status and worth are being questioned or denied, charges of pseudoscience rise and have more bite and bitterness. Let me illustrate this with a case from the past.

In 1844, writing anonymously, the Scottish publisher Robert Chambers authored a work on evolutionary theory, *Vestiges of the Natural History of Creation*. He argued that all organisms, especially humans, are the end results of a long, slow process of development from original primitive forms that themselves probably arose from inorganic matter. To make his case, Chambers drew eclectically but inventively from a wide variety of sources, grabbing bits of real science, of speculation, and of downright folklore. He writes about the nebular hypothesis, with its claim that the universe condensed naturally out of gas; about the lifelike forms of condensation on cold windowpanes ("frost ferns"); about insects appearing from an electric battery. We learn about the progressive nature of the fossil record; we hear of the German discoveries in embryology and the kind of recapitulation through which developing organisms travel; we are even invited to speculate that humans may not be the final result of evolution—that in the future a "crowning race" may supersede us. The book was a wild success— with the general public, that is. Benjamin Disraeli, future prime minister but a novelist in the 1840s and, as always, in tune with popular opinion, has one of his flightier characters—significantly, a woman—enthuse about it: "First there was nothing, then there was something; then, I forget the next, I think there were shells, then fishes; then we came, let me see, did we come next? Never mind that; we came at last. And the next change there will be something very superior to us, something with wings. Oh! that's it; we were fishes, and I believe we shall be crows." She continues, "It is impossible to contradict anything in it. You understand, it is all science; it is not like those books in which one says one thing and another the contrary, and both may be wrong. Everything is proved; by geology, you know" (Disraeli 1847, i, 225).

The British professional scientific community loathed *Vestiges* and, although the term had not caught on widely, branded it as pseudoscience (Secord 2000). Adam Sedgwick, professor of geology at the University of Cambridge, led the way. Having speculated that *Vestiges* could only have been written by a woman, he took back this suggestion as too vile to contemplate. It could never have been produced by a member of the fair sex: "The ascent up the hill of science is rugged and thorny, and ill fitted for the drapery of a petticoat." Rather, woman has "a soft and gentle temperament," "quick appreciation of character," and "instinctive knowledge of what is right and good" (Sedgwick 1845, 4). Sir David Brewster, a Scottish optician, biographer of Newton, and general man of science, wrote as follows: "Prophetic of infidel times, and indicating the unsoundness of general education, 'The Vestiges' has started into public favour with a fair chance of poisoning the fountains of science, and sapping the foundations of religion" (Brewster 1844, 471). He too worried about its popularity with women: "It would auger ill for the rising generation, if the mothers of England were infected with the errors of Phrenology. It would auger worse were they tainted with Materialism" (471). For him, the problem lay in the nature of woman. "The hold in which Providence has cast the female mind, does not present to us those phases of masculine strength which can sound depths, and grasp syllogisms, and cross examine nature" (471). And so it went. William Whewell brought together extracts from earlier works and published them (*Indications of the Creator*) without even mentioning the vile work. Sedgwick, having spent forty-five pages on the topic in mid-decade, returned to it in 1850, reprinting a thirty-five-page sermon from 1833, prefaced with five hundred pages of anti-*Vestiges* diatribe and three hundred pages of afterword in case the point had been missed. No wonder that the favorite of this group, the young Charles Darwin, already secretly an evolutionist, decided wisely to postpone his own publishing plans.

Why the tension? The reason was partly that *Vestiges*, as judged by the canons of good science, was in truth pretty dreadful. Frost ferns as evidence of life's spontaneous beginnings! Bugs coming out of a battery! Embryological upsets making for new species! Moreover, Chambers made no secret of his motivation, namely, to find something in the organic world that paralleled his belief in progress in the cultural and commercial world. Just as things are bound to improve in our daily lives thanks to our efforts, so things change for the better in the biological world. "A progression resembling development may be traced in human nature, both in the individual and in large groups of men. . . . Now all of this is in conformity with what we have seen of the progress of organic creation" (Chambers 1846, 400). He had no desire to satisfy any of the criteria of excellence that people like Whewell were articulating for quality science.

But there was more than this. At some level, these critical scientists were uneasy already (Ruse 1979). *Vestiges* exacerbated their problems and for this reason was deeply threatening. For a start, although the critics were all reformers who were trying to establish science as a profession, introducing science degrees into the older universities, founding and supporting organizations (especially the British Association for the Advancement of Science) that made science popular, they were trying to do it from within the establishment. They were university professors, contributors to important

journals, and officers of leading societies. They befriended Prince
Albert, the German, reform-minded consort of the Queen. They
were scientists, but they were also men of status—and religion was
a significant factor in their status. Fellows of Oxbridge colleges,
such as Sedgwick and Whewell, had to be ordained ministers in
the Anglican Church. In promoting science and insisting on its
greater role in British life, they were treading a fine line, and they
knew it. Change was acceptable only within certain limits. The
trouble was that many people thought that was impossible. This
was a time of radical philosophies that threatened massive revolu-
tionary change. It had already happened in France at the end of
the eighteenth century; it was to happen across Europe in 1848,
and it could happen in Britain, where the decade saw numerous
demonstrations and disputes from the Chartists and over the abo-
lition of the Corn Laws—laws that Cambridge professors favored
because they ensured high agricultural profits that benefited the
colleges to which they belonged. Were the scientific reformers not
part of this movement? They had to protest again and again that
they were not. And as Sedgwick in particular showed—always linking his hatred of evolution with his hatred of the philosophy of
progress—the attack on *Vestiges* was in large part a defense of his
own strategy of tempered change (Ruse 1996).

There was even more reason for tension. Evolution linked to
progress (implying that humans unaided can improve things)
threatened the reformers' own theology, which privileged the idea
that without God's love and help, we are lost. It threatened the
delicate balance they were trying to maintain between science and
religion, arguing that one can do science and yet keep up a good
theistic stance, admitting that miracles (as for the origins of species) occurred but in some sense were outside science. "The mystery of creation is not within the range of her legitimate territory;
she says nothing, but she points upwards" (Whewell 1837, 3:588).
Vestiges, arguing that all is law and rejecting miracles, threatened
this directly. This was the decade when British Protestants, especially Anglicans like Sedgwick and Whewell, were in turmoil over
the Oxford Movement, when the brightest stars (notably John

Henry Newman) were converting to Catholicism. Many conservatives, such as Dean Cockburn of York, were ready to criticize the scientists for embracing innovations like an old Earth and a restricted deluge. Hints that their position might open the way to evolution were anathema. Topping off their concerns was the division between professionals and the public. As professional scientists, the critics of *Vestiges* hated the presumption of the public to embrace what so clearly was unacceptable. They were denying the authority of the authorities! The status of women was also an issue. Women and their male supporters were pushing for better education—Tennyson wrote the *Princess* in 1847, a poem about a women's college, and Queen's College London, offering higher education for women, opened in the same year. For rather conservative people like the scientists, the whole *Vestiges* affair showed how dangerous this sort of thing could be.

Clearly, epistemology and criteria of scientific excellence are not all that count in disputes over the nature of fringe claims. This applies exactly to the Gaia controversy. Paradoxically, the supporters of evolutionary theory now represented the establishment, the professionals, and something outside was being labeled as pseudoscience. The reasons emerge from a quick review of the history of evolutionary theory. After Darwin published the *Origin*, people moved quickly to embrace evolution, but natural selection was ignored. Not until the 1930s, with the appearance of Mendelian genetics, did natural selection come into its own with the development of neo-Darwinism, or the synthetic theory (Ruse 2013a). By 1959, the hundredth anniversary of the *Origin*, biology had (and here the term seems appropriate) its paradigm. It was not entirely secure, because now molecular biology was attracting money, professors, and students, but it did exist and looked forward. Moreover, the next twenty years (until about the time of the Gaia controversy) saw major developments. Population genetics was enriched by molecular tools, such as gel electrophoresis, that revealed hitherto unobservable relationships. Major discoveries in paleontology—most notably Lucy (*Australopithecus afarensis*)—enriched our knowledge and understanding of the past, and the physical evidence was supplemented by new techniques, often involving computers. Biogeography benefited from plate tectonics, which confirmed hitherto implausible hypotheses about the drifting of continents and now-gone connections between their denizens. Systematics had a whole new approach, "cladistics," based on phylogenies. Studies of organic development began slowly but gathered steam as molecular biology began to uncover the ways in which genes spur growth. And above all, the study of social behavior was transformed as new models were introduced—notably through the work of theoreticians like William Hamilton and John Maynard Smith—and empirical studies confirmed their explanatory relevance and power.

But by 1980, the evolutionists were not happy campers (Hull 1988; Segerstrale 2000; Sepkoski and Ruse 2009). Evolutionary biologists were tearing themselves apart. They disagreed over population genetics, since new techniques had led to new results but to even greater differences of interpretation. We have seen how paleontologists were challenging strict Darwinism—which was the creed of Hamilton, Maynard Smith, and their supporters— and how these students of the fossils wanted alternative modes of thinking. The biogeographers differed among themselves, but this was nothing compared to the disagreements among systematists. The tensions between the cladists and more traditional workers were great and divisive. The fights over behavior capped everything; those who thought that we could now subsume human social activities under the Darwinian banner were accused of totally abrogating the standards of good science. The squabbles could get personal and nasty. Richard Lewontin had this to say about Edward O. Wilson, his colleague at Harvard: "I don't really think we are engaged primarily in an intellectual issue. I do not think that what he has been doing for the last ten years has been primarily motivated by a genuine desire to find out something true about the world, and therefore I don't think it is serious" (Segerstrale 1986, 75). Lewontin wrote a scathing review of a book on biology and culture coauthored by his fellow biologist (Lumsden and Wilson 1981). Lewontin explained himself as follows:

One of the reasons my book review . . . had a kind of sneering tone is that it is the way I genuinely feel about the project, namely, that it is not a serious, intellectual project. Because I have only two possibilities open to me. Either it is a serious intellectual project, and Ed Wilson can't think, or he can think, but it is not a serious project and therefore he is making all the mistakes he can—he does. If it is a really deep serious project, then he simply lowers himself in my opinion as an intellectual. (Segerstrale 1986, 75)

He continued: "I have to say that my chief feeling—I'll be honest about my chief feeling when I consider all this stuff—it's one of disdain. I don't know what to say, I mean, it's cheap!" (75).

We learned that one of Wilson's more audacious acts in *Sociobiology* was to extend his theorizing to humankind. It was this, above all, that so upset Lewontin, a recent convert to Marxism with the fervor that has marked converts from the days of Saint Paul. Ironically, given how Wilson was truly thinking, Lewontin's criticism, colored by his new emergentist philosophy, centered on Wilson's supposed undue fondness for reductionistic thinking. Wilson's critics saw him as arguing for a harsh view of the evolutionary process, not ameliorated by feelings of group based sentiment. Obviously, this was a textbook example of the gap between observation and interpretation. We know that, far from promoting extreme social Darwinism, Wilson was a thoroughgoing Spencerian holist. But, regardless of whether or not positions were properly attributed in this case, the point is that there were tensions. And they went the other way too.

Stephen Jay Gould was much more overtly a holist with an edge against individual selection. He was arguing for a "fuller" Darwinism, with activity at all levels—genes, organisms, species, and higher taxa. He was proposing his own theory of punctuated equilibrium in opposition to "Darwinian gradualism." He was attacking the central Darwinian notion of adaptation. In 1980, in a notorious article, he had declared that the Darwin-Mendel synthesis theory of evolution was "dead." Like Lovelock, he ventured into popular science and found that his tactic backfired when his fellow professionals judged his work inadequate. Even fifteen years later, the doyen of Darwinian evolutionists, John Maynard Smith, was responding bitterly.

> Gould occupies a rather curious position, particularly on his side of the Atlantic. Because of the excellence of his essays, he has come to be seen by non-biologists as the preeminent evolutionary theorist. In contrast, the evolutionary biologists with whom I have discussed his work tend to see him as a man whose ideas are so confused as to be hardly worth bothering with, but as one who should not be publicly criticized because he is at least on our side against the creationists. All this would not matter, were it not that he is giving non-biologists a largely false picture of the state of evolutionary theory. (Maynard Smith 1995, 46)

If this were not enough, assaults came from outside also. By the late 1970s, many (especially on the humanities' side of university campuses) were launching systematic attacks on the whole edifice of science. This opposition was foreshadowed by *Silent Spring*; radicalized by opposition to the military-industrial complex (which had played a major role in American foreign policy, most notably in the war in Vietnam); fortified by various philosophies that stressed the nonobjective nature of science (Michel Foucault's work was becoming increasingly influential, especially his thesis that knowledge was power); and inspired by various social movements, notably the feminist drive for respect and equality, science as a whole was under pressure. *Higher Superstition: The Academic Left and Its Quarrels with Science*, by Paul Gross and Norman Levitt (1994), defends science against what the authors took to be a barrage of unfair criticisms. It is not a great book (Ruse 1995), but despite, or perhaps because of, its shrill tone and selective array of targets, it does capture the spirit of the times.

Evolutionary biology was a favorite target of the critics, who (seizing on such metaphors as "selfish" genes) argued that in many respects it offered support for capitalism, for the sanction of violence, and especially for the repression of women. People were digging up some of the (by today's standards, rather appalling) views that Darwin had on the subject; he categorized females as immature males, not suited for the rigors of scholarship or business life, and more useful as the sustainers of family and feelings. (The protégé of Whewell and Sedgwick had learnt his lesson well.) Evolutionary theory, especially Darwinian theory focusing on adaptation, was under attack (Ruse 1982).

Paradoxically, paralleling this attack from the more radical elements of society, another attack came from the conservative evangelical Christians (Ruse 1988). Biblical literalists, calling themselves creationists or (to stress their credentials and intentions) creation scientists, argued vehemently that if evolution could be taught in schools, then a Genesis-based view of life's history—six-thousand year Earth history, divine creation of organisms, humans last, universal Flood—ought also to be taught. The original creationist movement received a major setback when a law mandating the "balanced treatment" in schools of evolution and creationism was defeated in Arkansas in 1981, but at the time, it was not pleasant to be an evolutionist in America. The recently elected president almost certainly sympathized with the creationist movement, and it was not obvious that evolution was emerging unscathed. Today, the creationists and their successors, the so-called intelligent design theorists, are doing all in their power to convince people that Darwin's thinking leads directly to national socialism. The last thing evolutionists needed at the time was the suggestion that Darwinism led in a straight line to eco-this and eco-that, to feminism in its extreme forms, and certainly not to Paganism— even though a few of these professionals were not that keen on pure Darwinism.

So when Gaia came bumbling in, it was seen as not just wrong but radically upsetting. Circling around to where we began, here again is John Postgate, microbiologist Fellow of the Royal Society:

As every thinking scientist knows, science has fallen badly into disrepute in the past couple of decades. The reasons, which

Chapter Eight

are multiple and very serious, are not relevant just now. The important thing is the consequence of that disrepute: fringe science, pseudoscience, obscurantism, wishful thinking and mysticism today find almost mediaeval favour even among educated people. For example who, a few decades ago, would have expected the surge of astrology, fringe medicine, faith healing, nutritional eccentricities, religious mysticism and a thousand other fads and cults which now plague developed societies? (Postgate 1988, 80)

Gaia is part of this threat to genuine science. "The ideas underlying Gaia have a proper basis in real science and her acolytes often offer common sense about environmental matters. But, dignified as a 'theory' and too often wrapped in mystical, cultish language— 'Gaia tells us so-and-so . . .'—such revealed wisdom, sound or not, is the antithesis of science" (80). What made things worse was that the betrayal came from within. Lovelock's scientific achievements were "tremendous and deserve our respect," but as for Gaia, "Let it remain a metaphor" (80).

For the Darwinian purists, it was appalling that the Gaia hypothesis did not so much as crash through the individual-group barrier but blithely ignored it. It was frightening that it used teleological language of a kind that had long been expelled from regular (that is, professional) science. Above all, it was associated with social and religious beliefs that were anathema to all evolutionists. These criticisms were true for scientists from Maynard Smith through the spectrum to Gould. Gould may have upset Maynard Smith with his attack on adaptationism, but he was no less keen that his science be considered professional. Much of Gould's maverick thinking was intended precisely to gain respect from professional scientists for his area of evolution, paleontology. In addition (as Maynard Smith acknowledged), Gould had been one of the major fighters against creationism. He knew a pseudoscience when he saw one. Here, then, we find the final piece of the puzzle. The professional community turned on Gaia not only because of its failings, but because of their own insecurities. John Maynard Smith was blunt about all of this. "Look Jim, all the trouble with Gaia is that we've had such agony with vitalism and group selection, and all these other things, and we thought we had it all worked out, and then you came along. You couldn't have chosen a worse moment" (DS).

Lovelock comments in response, "This, I think, explains a lot of the hostility" (DS). He's right. Gaia got caught in the crossfire, and when we recognize this, the final piece of our puzzle falls into place. The public adulation and the professional hostility were closely connected. Our history has shown us why and how.

ENVOI

Early in the year 2011, Jim Lovelock's two most devoted disciples, Timothy Lenton (who had worked with William Hamilton on the DMS cycle) and Andrew Watson (who coauthored the first Daisyworld paper with Lovelock), published a book on Earth history. Naturally enough *Revolutions That Made the Earth*, although not a book devoted exclusively to the Earth-as-organism hypothesis, made mention of the idea with a chapter entitled "Playing Gaia" about Daisyworld models and so forth. A few weeks later, the weekly news magazine *The Economist* published a short article on whether human interference with the workings of our globe is reason to argue that we have now entered a new geological epoch, the Anthropocene—"the recent age of man." Included in the discussion was a short (and accurate) account of Lenton and Watson's book, which was praised as "fascinating." The concept of Gaia was not mentioned. Does this omission tell us that interest in Gaia had vanished?

It seems fair to say that, by the end of the twentieth century, Gaia had achieved some degree of scientific acceptance. Tim Lenton had a major article on Gaia in *Nature* in 1998, discussing in some detail the nature of feedback systems. An article in *Science* on a Gaia conference carried the sympathetic title, "No Longer Willful, Gaia Becomes Respectable" (Kerr 1998, 393). This phrase plays on the fact that over the years Lovelock and his supporters had striven to cleanse Gaia of its crudest teleological excesses and other offensive features. Generally, critics no longer felt the need to cry "pseudoscience." In 2003, on the basis of groundbreaking studies of the oceans and their composition, Andrew Watson was elected to the Royal Society, so he must have been beyond any negative association with a disreputable idea. This change in attitude toward Lovelock's worldview was due not only to the efforts of Gaia enthusiasts but also to the fact that the status of evolutionary biology was much more secure than it had been twenty years previously. Painstaking empirical and theoretical work had ended conflicts over such subjects as classification and social behavior, and people were moving ahead strongly in these and related areas, often making full use of the findings and techniques of molecular biology. John Maynard Smith, for instance, once a great critic of Gaia, now felt sufficiently reconciled that he spent a night with Lovelock and his wife Sandy at their home in Devon.

Yet, as the *Science* article made clear, this did not mean that everyone was now coming on board and agreeing that Gaia, in whatever form, was now good and fruitful science. Lovelock was being praised for his insistence that life is important in the overall history of the planet and for his thinking about cycles and feedback systems. A year or two later, in an editorial in *Science* entitled "Earth System Science" (the science of the state of and processes on Earth), J. Lawton wrote, "James Lovelock's penetrating insights that a planet with abundant life will have an atmosphere shifted into extreme dynamic disequilibrium, and that Earth is habitable because of complex linkages and feedbacks between the atmosphere, oceans, lands, and biosphere, were major stepping-stones in the emergence of this new science" (Lawton 2001, p. 1965).

But some people were still stuck on the old concern, expressed recently by Hamilton, that Gaia was claiming that in some sense the various feedbacks all add up to a controlled system that is good for the planet and for the life on it. For all the effort, there was still a purpose-driven flavor that was unacceptable. The old Dawkins criticism still carried weight: we just don't find final cause in biology except in the naturalized form that Darwin's natural selection makes possible. And Earth is a unique instance, not the end result of a selective process. There are feedbacks, certainly, but there is nothing to guarantee that they promote some kind of stability. Andy Watson, of all people, was now agreeing that the feedback involved in the DMSP cycle described by Lovelock and others at the end of the 1980s was probably positive rather than negative. Others were (and are still) making related criticisms. The American biologist Tyler Volk argues that waste products are the key: "The wastes from creatures in one biochemical guild can be nutrients to those in another guild. CO_2 is waste we expel, but it is airborne food to green plants. The nitrogen gas N_2 is waste from the denitrifying bacteria, but a nutrient to the nitrogen-fixing bacteria" (Volk 1998, 34). He stresses that the wastes "are not produced at cost to give to other creatures." Rather, "The wastes are simply by-products. The biosphere is a stupendous network of waste by-products that are also nutrients" (38). If true, this is quite remarkable, but ultimately (as Andrew Watson allows), it had no inherent meaning, nor could it guarantee stability. "It's just chance." We are here, so it happened, but, given the number of solar systems overall—"someone throws dice 10^{22} times"—what else would you expect? (AW)

In addition, there was little enthusiasm for Daisyworld models. Robert May dismissed them as "a marginal note on a more professional enterprise." Richard Dawkins said that Daisyworld produces "an illusion of control" (Lewin 1992, 117). This seems rather harsh, and Tim Lenton in particular has persevered in showing their promise (Lenton 1998). The basic problem seems to be less with the mathematics and more with the applicability. Consider the black daisies, which thrive as the planet gets hotter. They do well because the temperature is right for them, but eventually, as a kind of by-product, they make the planet too hot for their own good, and the white daisies start to move in and grow. The proliferation of white daisies begins to cool the planet down, and at some point a kind of equilibrium is achieved. But what would happen in real life? The temperature of the planet is not the result of any intention. Reproduction is what counts. If the success of the black daisies ultimately leads to conditions that no longer favor them, allowing the success of competitors, then selection pressure would favor any new variation (e.g., greater tolerance for heat) that keeps the black daisies thriving. Although it is true that the daisies increase or decrease because of individual-selection forces, selfish genes might eventually wipe everything out or at least lead to a less than ideal situation.

Overall, therefore, apart from a small group of supporters, and despite general agreement that Gaia had pushed thinking in some important directions, by around 2000 the Gaia hypothesis as professional science had lost momentum. "Jim never came up with what was for many people the convincing mechanism [of homeostasis], and that was the kind of Holy Grail they were looking for" (TL). Part of the problem, of course, was that Gaia did not have much of an institutional base. Charles Darwin tucked himself away in the wilds of Kent, but he managed to build up a group of devoted supporters—Thomas Henry Huxley and Joseph Hooker in Britain, Asa Gray in America, Ernst Haeckel in Germany (Ruse 1979). Lovelock tucked himself away in the wilds of Devon, but he lacks enough supporters to create momentum, even if that were possible. If you are rude about academia but lack Darwin's uncanny ability to flatter and cajole, to take what you can and compromise on the rest for the moment, you cannot establish a beachhead. Unfortunately, "having invested his entire life in this idea," Lovelock tends to be "quite tribal about Gaia" (AW). He was incensed when Watson and Lenton organized a conference and (at the urging of the sponsors) spoke only of coevolution and not explicitly of Gaia. Watson says that Lovelock "accused me of reneging on Gaia in order to advance my research career," and defends himself by saying, "We're trying to organize a meeting. We want people to come. If we call it Gaia, we will get the committed Gaians. But there's an awful lot of people won't come, because they think it's a bunch of airheads" (AW). In the end, Lovelock gave a keynote speech, but Gaia was not included in the conference title.

Lovelock turned from straight science to issuing apocalyptic warnings about the future fate of our planet and to pointing out the severe damage we are doing to our planet through fuel consumption and agricultural practices (Lovelock 2006). Recently, he argued that things are almost beyond control, that the planet is on its way to much hotter temperatures, and at best but a few humans will survive. "Our planet looks after itself. All we can do is try to save ourselves" (Lovelock 2009, 13). Fortunately, "exposed to the severe selection pressures soon to come, we, as a species, may grow up and become capable" (234). Even more recently Lovelock has done an about face, rejecting such extremism. Ever the maverick, he now refers to much of the global warming movement as "green drivel." Seeing his brainchild, Gaia, slipping away as the world approaches a boiling Armageddon, he now sharply criticizes the prophets of doom, agreeing that while global warming is occurring, it is not happening as rapidly as they say. Organisms fight back. Characteristically, Lovelock praises himself for having the courage to change his mind, pointing out with satisfaction that if he were not an independent scientist, he would not dare to speak out against the norm (Johnston 2012; Hickman 2012).

Either way, such talk makes his scientific supporters queasy. They do share his vision—"It's a fantastic jewel of a planet" (AW)—and his sense that "there's a moral dimension" (TL) to the discussion, but they want the science kept separate and developed in its own right. They fear that Lovelock seems to have pulled back from the drives that helped to win at least some modicum of respectability for Gaia. His use of the language of organism has become more literal and less metaphorical. Somewhat ruefully, Lenton comments, "Earlier on in my career [the 1990s], I was more with Jim on my use of the Gaia banner or word." Now he is more cautious. "Metaphors are useful to a point. 'Earth is alive' or 'Earth is an organism' are useful for popular uses or consumption, but are not helpful when talking to other scientists" (TL). His caution is wise, given that, in Lovelock's recent writings, the teleological underpinning has surfaced again. "I call Gaia a physiological system because it appears to have the unconscious goal of regulating the climate and the chemistry at a comfortable state for life" (Lovelock 2006, 15). He continues in the same vein. "Gaia regulates its temperature at what appears to be optimal for whatever life is inhabiting it" (45). Even the details are enough to make a conventional biologist cringe. "Why do we pee? Not so silly a question as it might seem. The need to rid oneself of waste products like salt, urea, creatinine, and numerous other scraps of metabolism is obvious but only part of the answer. Perhaps we pee for altruistic reasons. If we and other animals did not pass urine some of the vegetable life of the Earth might be starved of nitrogen" (19). Along with this, a more mystical, Quaker-inspired philosophy is coming to the fore. "It is time, I think, that theologians shared with scientists their wonderful word, 'ineffable'; a word that expresses the thought that God is immanent but unknowable" (138). Lovelock now speaks warmly of deep ecologists and even has a kind word for New Age thinkers! (147)

Lynn Margulis's enthusiasms later in her life did little to help the standing of Gaia. Increasingly, she was seen as beyond the bounds of acceptable professional science. Her views on the origins of sex (seeing it as a form of symbiosis) were considered bad enough. Even worse was her suggestion that spirochetes (spiral-shaped bacteria such as the one that causes syphilis) are symbiotically involved in the cause of AIDS. And truly dreadful was her endorsement (to the extent of getting a paper promoting the idea placed in the *Proceedings of the National Academy of Sciences*) of the idea that butterflies are the product of the hybridization of certain extinct species of insect and velvet worms (Williamson 2009; see also Hart and Grosberg 2009). (Velvet worms are caterpillar-like organisms of the southern hemisphere that have peculiar mating habits. The male deposits sperm in containers on its head, and then puts the whole thing into the vagina of the female.) One expert in the field commented that the paper was better suited for the "*National Enquirer* rather than the National Academy." The refereeing system was at once made much more stringent.

Is Gaia a "failed paradigm"? Andrew Watson remains loyal: "No! It's part of the Earth systems revolution." Fair enough, but he adds, "The vision of the Earth as a single organism doesn't quite work" (AW). So in the realm of science, in one sense (what many would consider the lasting and important sense), there is real success. But, focusing just on the metaphor, it is hard to escape negative judgment, although failures in the history of science can be as interesting as successes, and perhaps even more revealing about science and scientists. Yet professional science is but part of the story. Lovelock and Margulis are revered in many quarters as the couple who put the vision on the map, who gave a philosophy a firm scientific basis. Gaia was and continues to be a public phenomenon, from the philosopher Mary Midgley to the truly irrepressible Oberon Zell-Ravenheart, who, inspired by the Harry Potter book and film series, has recently started his own online academy, the Grey School of Wizardry (Zell-Ravenheart 2004; 2006), including coursework on social intercourse with the moon, alchemy, beast mastery (totems and familiars), and mathemagicks (we start with Pythagoras). Frankly, it is hard not to crack a smile at this point; indeed, I would not altogether trust a person who did not. But after you discard the trimmings (Is there anything here truly weirder than sacred underpants, or flawless red heifers, or the many other manifestations of the American religious spirit?) ultimately, Zell-Ravenheart's philosophy is one version of the almost mystical feeling that so many people have about our planet— the feeling that the world where we spend our days trying to earn a living, where we interact with others in love and friendship (and too often with animosity), trying to draw meaning and purpose from our sojourn here on Earth, must be more than just blind molecules in motion. Plato was right and the atomists were wrong. Our planet is alive. This is so, whether you think that humans are the ultimate purpose or—as Pagans and other Gaia enthusiasts think—that the ultimate purpose lies somehow in Mother Earth herself.

Failure as science is balanced by success as philosophy. We see why the project to which Lovelock and Margulis committed themselves ultimately could not work. Conceal and modify it as you may, if you are true to the Gaia project—to see the world as an organism—you are committed to something deeply end-directed, deeply teleological. Earth is in balance because that is its destiny, that is what it should be. That is a good thing. Modern science, shorn of teleology and value, simply will not allow this kind of vision within its borders. Hence, the Gaia hypothesis was doomed to failure. We also see why the project to which Lovelock and Margulis committed themselves worked wonderfully. For many people, the vision endures and in their books—in this book—Jim Lovelock and Lynn Margulis are heroes for their attempt to meld the two worlds of the 1960s. They brought together the world of science and technology; the world of space travel and related achievements (Lovelock to this day upsets the Greens with his proselytizing on behalf of nuclear energy); and the world of feeling, of mysticism and religion, of New Age movements and liberating philosophies of life that are "more prescient than the scientists" (Lovelock 2006, 147). So no more talk of failure. Lovelock and Margulis were big people with a big vision. Whether science likes it or not, the vision lives on.

"It is most certainly an organism—and alive!" (JL)

References

Adler, M. 2006. *Drawing Down the Moon: Witches, Druids, Goddess-Worshipers and other Pagans in America.* Rev. ed. with expanded appendix. New York: Penguin.

Agassiz, E. C., ed. 1885. *Louis Agassiz: His Life and Correspondence.* 2 vols. Boston: Houghton Mifflin.

Allee, W. C. 1927. Animal aggregations. *Quarterly Review of Biology* 2: 367–98.

Allee, W., A. Emerson, T. Park, O. Park, and K. Schmidt. 1949. *Principles of Animal Ecology.* Philadelphia, PA: Saunders.

Allen, P. G. 1990. The woman I love is a planet; the planet I love is a tree. In *Reweaving the World: The Emergence of Ecofeminism,* edited by I. Diamond and G. F. Orenstein, 52–57. San Francisco: Sierra Club.

Anonymous. 2011. A man-made world. *Economist,* May 23, 81–83.

Armstrong, A. H. 1940. *The Architecture of the Intelligible Universe in the Philosophy of Plotinus.* Cambridge: Cambridge University Press.

Barnes, J., ed. 1984. *The Complete Works of Aristotle.* 2 vols. Princeton, NJ: Princeton University Press.

Beierwaltes, W. 2003. Plato's *Timaeus* in German idealism: Schelling and Windischmann. In *Plato's Timaeus as Cultural Icon,* edited by G. J. Reydams-Schils, 267–90. Notre Dame, IN: University of Notre Dame Press.

Benison, S., A. C. Barger, and E. L. Wolfe. 1987. *Walter B. Cannon: The Life and Times of a Young Scientist.* Cambridge, MA: Harvard University Press.

Bergson, H. 1907. *L'évolution créatrice.* Paris: Alcan.

Bernstein, J. 1982. *Science Observed.* New York: Basic Books.

Berry, T. 2009. *The Sacred Universe: Earth, Spirituality, and Religion in the Twenty-first Century.* New York: Columbia University Press.

Bowler, P. J. 1989. *The Mendelian Revolution: The Emergence of Hereditarian Concepts in Modern Science and Society.* London: Athlone Press.

Boyle, R. (1688) 1966. A disquisition about the final causes of natural things. In *The Works of Robert Boyle*, edited by T. Birch, 5:392–444. Hildesheim, Germany: Georg Olms.

———. 1996. *A Free Enquiry into the Vulgarly Received Notion of Nature.* Edited by E. B. Davis, and M. Hunter. Cambridge: Cambridge University Press.

Brewster, D. 1844. Vestiges. *North British Review* 3:470–515.

Brockman, J. 1995. *The Third Culture: Beyond the Scientific Revolution.* New York: Simon and Schuster.

Browne, J. 1995. *Charles Darwin: Voyaging.* Vol. 1 of *A Biography.* New York: Knopf.

Burtt, E. A. 1932. *The Metaphysical Foundations of Modern Physical Science.* 2nd ed. New York: Harcourt, Brace.

Caldeira, K. 1989. Evolutionary pressures on plankton production of atmospheric sulphur. *Nature* 337:732–34.

Cannon, W. B. 1931. *The Wisdom of the Body.* Cambridge, MA: Harvard University Press.

Capra, F. 1975. *The Tao of Physics: An Exploration of the Parallels between Modern Physics and Eastern Mysticism.* Berkeley, CA: Shambhala.

———. 1982. *The Turning Point: Science, Society, and the Rising Culture.* New York: Simon and Schuster.

———. 1995. Deep ecology: A new paradigm. In *Deep Ecology for the 21st Century*, edited by G. Sessions, 19–25. Boston: Shambhala. Reprinted from *Earth Island Journal* 2 (1987).

Carey, J. 2009. *William Golding: The Man Who Wrote Lord of the Flies.* New York: Free Press.

Carson, R. L. 1941. *Under the Sea Wind: A Naturalist's Picture of Ocean Life.* New York: Simon and Schuster.

———. 1951. *The Sea Around Us.* New York: Oxford University Press.

———. 1955. *The Edge of the Sea.* Boston: Houghton Mifflin.

———. 1962. *Silent Spring.* New York: Houghton Mifflin.

Chambers, R. 1844. *Vestiges of the Natural History of Creation.* London: J. Churchill.

———. 1846. *Vestiges of the Natural History of Creation.* 5th ed. London: J. Churchill.

Charlson, R. J., J. E. Lovelock, M. O. Andreae, and S. J. Warren. 1987. Oceanic phytoplankton, atmospheric sulphur, cloud albedo and climate. *Nature* 326:655–61.

Clarke, B. 2012. Gaia is not an organism: Scenes from the early scientific collaboration between James Lovelock and Lynn Margulis. In *Lynn Margulis: The Life and Legacy of a Scientific Rebel*, edited by D. Sagan, 32–43. White River Junction, VT: Chelsea Green.

Coleridge, S. T. 1848. *Hints Towards the Formation of a More Comprehensive Theory of Life.* Edited by S. B. Watson. London: J. Churchill.

Cooper, J. M., ed. 1997. *Plato: Complete Works.* Indianapolis, IN: Hackett.

Cranston, S. L. 1994. *HPB: The Extraordinary Life and Influence of Helena Blavatsky, Founder of the Modern Theosophical Movement.* New York: Putnam.

Darwin, C. 1859. *On the Origin of Species by Means of Natural Selection, or the Preservation of Favoured Races in the Struggle for Life.* London: John Murray.

———. 1861. *Origin of Species.* 3rd ed. London: John Murray.

———. 1959. *The Origin of Species by Charles Darwin: A Variorum Text.* Edited by M. Peckham. Philadelphia: University of Pennsylvania Press.

———. 1969. *Autobiography.* New York: W. W. Norton.

———. 1975. *Charles Darwin's Natural Selection, Being the Second Part of His Big Species Book Written from 1856 to 1858.* Edited by R. C. Stauffer. Cambridge: University of Cambridge Press.

———. (1860) 1985–. *The Correspondence of Charles Darwin.* Vol. 8. Cambridge: Cambridge University Press.

Dawkins, R. 1976. *The Selfish Gene.* Oxford: Oxford University Press.

———. 1982. *The Extended Phenotype: The Gene as the Unit of Selection.* Oxford: W. H. Freeman.

———. 1995. *A River Out of Eden.* New York: Basic Books.

Descartes, R. (1644) 1955. *The Principles of Philosophy.* Vol. 1 of *The Philosophical Works of Descartes.* 2 volumes. Translated by E. Haldane, and G. R. T. Ross, 1:201–302. New York: Dover.

———. 1964. *Discourse on Method.* In *Philosophical Essays,* 1–57. Translated by Laurence Lafleur. Indianapolis, IN: Bobbs-Merrill.

Dijksterhuis, E. J. 1961. *The Mechanization of the World Picture.* Oxford: Oxford University Press.

Disraeli, B. 1847. *Tancred, or the New Crusade.* 3 vols. London: Henry Colburn.

Doolittle, W. F. 1981. Is nature really motherly? *CoEvolution* 29:58–62.

Driesch, H. 1908. *The Science and Philosophy of the Organism.* London: Black.

Dronke, P., ed. 1988. *A History of Twelfth-Century Western Philosophy.* Cambridge: Cambridge University Press.

Duhem, P. (1906) 1954. *The Aim and Structure of Physical Theory.* Foreword by Prince Louis de Broglie; translated by Philip P. Wiener. Princeton, NJ: Princeton University Press.

Duncan, D., ed. 1908. *Life and Letters of Herbert Spencer.* London: Williams and Norgate.

Easton, S. C. 1980. *Rudolf Steiner: Herald of a New Epoch.* Hudson, NY: Anthroposophic Press.

Ehrlich, P. R. 1991. Coevolution and its applicability to the Gaia hypothesis. In *Scientists Debate Gaia*, edited by S. H. Schneider and P. J. Boston, 19–22. Cambridge, MA: MIT Press.

Eldredge, N., and S. J. Gould. 1972. Punctuated equilibria: An alternative to phyletic gradualism. In *Models in Paleobiology*, edited by T. J. M. Schopf, 82–115. San Francisco: Freeman, Cooper.

Emerson, A. E. 1939. Social coordination and the superorganism. *American Midland Naturalist* 21:182–209.

Emerson, R. W. 1836. *Nature*. Boston: James Munroe.

Evangelical Lutheran Church in America. 1996. Basis for our caring. In *This Sacred Earth: Religion, Nature, Environment*, edited by R. S. Gottlieb. 243–50. New York: Routledge.

Evans, M. A., and H. E. Evans. 1970. *William Morton Wheeler, Biologist*. Cambridge, MA: Harvard University Press.

Farlow, J. O., C. V. Thompson, and D. E. Rosner. 1976. Plates of the dinosaur Stegosaurus: Forced convection heat loss fins? *Science* 192:1123–25.

Fenchel, T. 2003. *The Origin and Early Evolution of Life*. New York: Oxford University Press.

Freeman, S., and J. C. Herron. 2004. *Evolutionary Analysis*. 3rd ed. Englewood Cliffs, NJ: Prentice-Hall.

Garber, D. 1992. *Descartes' Metaphysical Physics*. Chicago: University of Chicago Press.

Golding, W. 1954. *Lord of the Flies*. London: Faber and Faber.

———. 1956. *Pincher Martin*. London: Faber and Faber.

———. 1959. *Free Fall*. London: Faber and Faber

———. 1980. *Rites of Passage*. London: Faber and Faber.

———. 1983. "Nobel Lecture, 7 December 1983." Available online at www .nobelprize.org/nobelprizes/literature.

———. 1987. *Close Quarters*. London: Faber and Faber.

———. 1989. *Fire Down Below*. London: Faber and Faber.

Gordin, M. 2012. *The Pseudoscience Wars: Immanuel Velikovsky and the Birth of the Modern Fringe*. Chicago: University of Chicago Press.

Gould, S. J. 1980. Is a new and general theory of evolution emerging? *Paleobiology* 6:119–30.

———. 1987. Kropotkin was no crackpot. *Natural History* 106 (June): 12–21.

———. 1991. *Bully for Brontosaurus*. New York: W. W. Norton.

Gould, S. J., and R. C. Lewontin. 1979. The spandrels of San Marco and the Panglossian paradigm: A critique of the adaptationist programme. *Proceedings of the Royal Society of London*, ser. B: *Biological Sciences*, no. 205: 581–98.

Grant, R. B., and P. R. Grant. 1989. *Evolutionary Dynamics of a Natural Pop-*

ulation: The Large Cactus Finch of the Galápagos. Chicago: University of Chicago Press.

Griggs, D. 1939. A theory of mountain building. *American Journal of Science* 237:611–50.

Gross, P. R., and N. Levitt. 1994. *Higher Superstition: The Academic Left and Its Quarrels with Science.* Baltimore, MD: Johns Hopkins University Press.

Haeckel, E. 1895. *Monism as Connecting Religion and Science: The Confession of Faith of a Man of Science.* London: Adam and Charles Black.

Hall, A. R. 1954. *The Scientific Revolution, 1500–1800: The Formation of the Modern Scientific Attitude.* London: Longman, Green.

————. 1996. *Henry More and the Scientific Revolution.* Cambridge: Cambridge University Press.

Hamilton, W. D. 1964. The genetical evolution of social behaviour. *Journal of Theoretical Biology* 7:1–52.

————. 1998. Letter to the editor. *New Scientist,* June 27, 52.

Hamilton, W. D., and T. M. Lenton. 2005. Spora and Gaia: How microbes fly with their clouds. In *Narrow Roads of Gene Land.* Vol. 3. *Last Words,* edited by W. D. Hamilton, 271–89. Oxford: Oxford University Press. Reprinted from *Ethnology, Ecology, and Evolution* 10 (1998): 1–16.

Harding, S. 2006. *Animate Earth: Science, Intuition, and Gaia.* White River Junction, VT: Chelsea Green.

Harding, S., and L. Margulis. 2009. Water Gaia: Three and a half thousand million years of wetness on planet Earth. In *Gaia in Turmoil: Climate Change, Biodepletion, and Earth Ethics in an Age of Crisis,* edited by E. Crist and H. B. Rinker, 41–60. Cambridge, MA: MIT Press.

Hart, M. W., and R. K. Grosberg. 2009. Caterpillars did not evolve from onychophorans by hybridogenesis. *Proceedings of the National Academy of Sciences* 106:19906–9.

Harvey, P. 2004. Not only termites should be alarmed. (Review of *Acquiring Genomes,* by L. Margulis and D. Sagan). *Times Higher Educational Supplement,* April 23, no. 1637, 31.

Heinlein, R. 1961. *Stranger in a Strange Land.* New York: Putnam.

Hemenway, T. 2009. *Gaia's Garden, Second Edition: A Guide to Home-Scale Permaculture.* White River Junction, VT: Chelsea Green.

Henderson, L. J. 1913. *The Fitness of the Environment.* New York: Macmillan.

————. 1917. *The Order of Nature.* Cambridge, MA: Harvard University Press.

Hickman, L. 2012. James Lovelock: The UK should be going mad for fracking. *Guardian,* June 15.

Hölldobler, B., and E. O. Wilson. 2008. *The Superorganism: The Beauty, Elegance, and Strangeness of Insect Societies.* New York: W. W. Norton.

Hull, D. L. 1988. *Science as a Process.* Chicago: University of Chicago Press.

Hunt, H. A. K. 1976. *A Physical Interpretation of the Universe: The Doctrines of Zeno the Stoic.* Carlton, Australia: Melbourne University Press.

Hunt, L. 1998. Send in the clouds. *New Scientist,* May 30, 28–33.

Hutchinson, G. E. 1948. Circular causal systems in ecology. *Annals of the New York Academy of Science* 50:221–46.

———. 1970. Foreword to L. Margulis, *Origin of Eukaryotic Cells,* xvii. New Haven, CT: Yale University Press.

Hutton, J. 1788. Theory of the Earth; or, an investigation of the laws observable in the composition, dissolution, and restoration of land upon the globe. *Transactions of the Royal Society of Edinburgh* 1 (2): 209–304.

———. 1795. *Theory of the Earth, with Proofs and Illustrations.* Edinburgh: William Creech.

Hutton, R. 2001. *The Triumph of the Moon: The History of Modern Pagan Witchcraft.* Oxford: Oxford University Press.

Hynes, H. P. 1989. *The Recurring Silent Spring.* New York: Teachers College Press.

Jang, Hwee-Yong. 2007. *The Gaia Project: 2012; The Earth's Coming Great Changes.* Woodbury, MN: Llewellyn.

Johansen, T. K. 2004. *Plato's Natural Philosophy: A Study of the Timaeus-Critias.* Cambridge: Cambridge University Press.

Johnston, I. 2012. "Gaia" scientist James Lovelock: I was "alarmist" about climate change. MSNBC, *World News,* April 23.

Joseph, L. E. 1990. *Gaia: The Growth of an Idea.* New York: St. Martin's.

Kant, I. (1790) 2001. *Critique of the Power of Judgment.* Edited by P. Guyer. Cambridge: Cambridge University Press.

———. (1786) 2009. *Metaphysical Foundations of Natural Science.* Edited and translated by J. Bennett. Available only online at www.earlymoderntexts .com/pdf/kantmeta.pdf.

Keary, P., K. A. Klepeis, and F. J. Vine. 2009. *Global Tectonics.* 3rd ed. Hoboken, NJ: Wiley-Blackwell.

Keller, E. F. 1986. One woman and her theory. *New Scientist,* July 3, 46–50.

Kepler, J. 1977. *The Harmony of the World.* Translated by E. J. Aiton, A. M. Duncan, and J. V. Field. Philadelphia, PA: American Philosophical Society.

Kerr, R. A. 1998. No longer willful, Gaia becomes respectable: The Gaia hypothesis, that Earth is a single huge organism intentionally creating an optimum environment for itself, has been made more palatable; interesting science is coming of it. *Science* 240:393–95.

Kirchner, J. W. 1989. The Gaia hypothesis: Can it be tested? *Reviews of Geophysics* 87:223–35.

———. 1990. Gaia metaphor unfalsifiable (letter). *Nature* 345:470.

———. 1991. The Gaia hypotheses: Are they testable? Are they useful? In

Scientists on Gaia, edited by S. H. Schneider and P. J. Boston, 38–46. Cambridge, MA: MIT Press.

———. 2002. The Gaia hypothesis: Fact, theory, and wishful thinking. *Climatic Change* 52:391–408.

———. 2003. The Gaia hypothesis: Conjectures and refutations. *Climatic Change* 58:21–45.

Lachman, G. 2007. *Rudolf Steiner: An Introduction to His Life and Work.* New York: Penguin.

Lawton, J. 2001. Editorial: Earth system science. *Science* 292:1965.

Lear, L. 1997. *Rachel Carson: Witness for Nature.* New York: Henry Holt.

Lenton, T. M. 1998. Gaia and natural selection. *Nature* 394:439–47.

———. 2002. Testing Gaia: The effect of life on Earth's habitability and regulation. *Climatic Change* 52: 409–22.

Lenton, T. M., and A. Watson. 2011. *Revolutions That Made the Earth.* Oxford: Oxford University Press.

Leopold, A. 1949. *A Sand County Almanac.* New York: Oxford University Press.

———. 1979. Some fundamentals of conservation in the Southwest. *Environmental Ethics* 1:131–42.

Lewin, R. 1992. *Complexity: Life at the Edge of Chaos.* New York: Collier.

Longgood, W. 1960. *The Poisons in Your Food.* New York: Simon and Schuster.

Lovelock, J. E. 1972. Gaia as seen through the atmosphere. *Atmospheric Environment* 6:579–580.

———. 1979. *Gaia: A New Look at Life on Earth.* Oxford: Oxford University Press.

———. 1982. A physicist's world view (Review of F. Capra, *The Turning Point*). *New Scientist*, July 22, 254.

———. 1988. *Ages of Gaia: A Biography of Our Living Earth.* New York: W. W. Norton.

———. 1990. Hands up for the Gaia hypothesis. *Nature* 344:100–102.

———. 1991. *Healing Gaia: Practical Medicine for the Planet.* New York: Harmony.

———. 1998. Letter to editor. *New Scientist*, July 11, 57.

———. 2000. *Homage to Gaia.* Oxford: Oxford University Press.

———. 2006. *The Revenge of Gaia: Earth's Climate Crisis and the Fate of Humanity.* New York: Basic Books.

———. 2009. *The Vanishing Face of Gaia: A Final Warning.* New York: Basic Books.

Lovelock, J. E., and S. Epton. 1975. The quest for Gaia. *New Scientist*, February 6, 304–9.

Lovelock, J. E., and L. Margulis. 1974a. Homeostatic tendencies of the Earth's atmosphere. *Origins of Life* 5:93–103.

———. 1974b. Atmospheric homeostasis by and for the biosphere: The Gaia hypothesis. *Tellus* 26:1–10.

Lumsden, C. J., and E. O. Wilson. 1981. *Genes, Mind, and Culture.* Cambridge, MA: Harvard University Press.

Lyell, C. 1830–33. *Principles of Geology: Being an Attempt to Explain the Former Changes in the Earth's Surface by Reference to Causes now in Operation.* 3 volumes. London: John Murray.

Magendie, F. 1843. *An Elementary Treatise on Human Physiology.* New York: Harper.

Malthus, T. R. (1826) 1914. *An Essay on the Principle of Population.* 6th ed. London: Everyman.

Margulis, L. 1970. *The Origin of Eukaryotic Cells: Evidence and Research Implications for a Theory of the Origin and Evolution of Microbial, Plant, and Animal Cells on the Precambrian Earth.* New Haven, CT: Yale University Press.

———. 1999. *Symbiotic Planet: A New Look at Evolution.* New York: Basic Books.

———. 2002. *Acquiring Genomes: The Theory of the Origins of the Species.* New York: Basic Books.

———. 2005. Hans Ris (1914–2004): Genophore, chromosomes and the bacterial origin of chloroplasts. *International Microbiology* 8:145–48.

———, and J. E. Lovelock. 1974. Biological modulation of the Earth's atmosphere. *Icarus* 21:471–89.

Margulis, L., and D. Sagan. 1997. *Slanted Truths: Essays on Gaia, Symbiosis and Evolution.* Secaucus, NJ: Copernicus Books.

Maturana, H. R., and F. J. Varela. 1980. *Autopoiesis and Cognition: The Realization of the Living.* Dordrecht, Holland: Reidel.

Maynard Smith, J. 1978. *The Evolution of Sex.* Cambridge: Cambridge University Press.

———. 1982. *Evolution and the Theory of Games.* Cambridge: Cambridge University Press.

———. 1995. Genes, memes, and minds. *New York Review of Books* 42 (19): 46–48.

Mayr, O. 1989. *Authority, Liberty, and Automatic Machinery in Early Modern Europe.* Baltimore, MD: Johns Hopkins University Press.

McMullin, E. 1983. Values in science. In *PSA 1982*, edited by P. D. Asquith and T. Nickles, 3–28. East Lansing, MI: Philosophy of Science Association.

Meine, C. 1988. *Aldo Leopold: His Life and Work.* Madison: University of Wisconsin Press.

Menand, L. 2001. *The Metaphysical Club: A Story of Ideas in America.* New York: Farrar, Straus, and Giroux.

Merchant, C. 1995. *Earthcare: Women and the Environment*. London: Routledge.

Mervis, J. 2009. An inside/outside view of US science. *Science* 325:132–33.

Midgley, M. 1979. Gene-juggling. *Philosophy* 54:439–58.

———. 2005a. Gods and goddesses. In *The Essential Mary Midgley*, edited by D. Midgley, 359–66. London: Routledge.

———. 2005b. The unity of life. In *The Essential Mary Midgley*, edited by D. Midgley, 373–78. London: Routledge.

———, ed. 2007. *Earthy Realism: The Meaning of Gaia*. Exeter, UK: Imprint Academic.

Mitman, G. 1992. *The State of Nature: Ecology, Community, and American Social Thought, 1900–1950*. Chicago: University of Chicago Press.

Muir, J. 1966. *John of the Mountains: The Unpublished Journals of John Muir*. Edited by L. M. Wolfe. Madison: University of Wisconsin Press.

Næss, A. 1995a. The place of joy in a world of fact. In *Deep Ecology for the 21st Century*, edited by G. Sessions, 249–58. Boston: Shambhala. Reprinted from the *North American Review* 258 (1973).

———. 1995b. The deep ecological movement: Some philosophical aspects. In *Deep Ecology for the 21st Century*, edited by G. Sessions, 64–84. Boston: Shambhala. Reprinted from *Philosophical Inquiry* 8 (1986).

———. 1995c. Self-realization: An ecological approach to being in the world. In *Deep Ecology for the 21st Century*, edited by G. Sessions, 225–39. Boston: Shambhala.

Nagel, E. 1961. *The Structure of Science: Problems in the Logic of Scientific Explanation*. New York: Harcourt, Brace and World.

Neumann, H.-P. 2010. Machina Machinarum: Die Uhr als Begriff und Metapher zwischen 1450 und 1750. *Early Science and Medicine* 15:122–91.

Newton, J. L. 2006. *Aldo Leopold's Odyssey: Rediscovering the Author of "A Sand County Almanac."* Washington, DC: Island Press.

Norwood, V. 1993. *American Women and Nature*. Chapel Hill: University of North Carolina Press.

Odum, E. P. 1969. The strategy of ecosystem development. *Science* 164: 262–70.

———. 1989. *Ecology and Our Endangered Life Support System*. Sunderland, MA: Sinauer.

Olby, R. C. 1963. Charles Darwin's manuscript of Pangenesis. *British Journal for the History of Science* 1 (3): 251–63.

Oldroyd, D. R. 1996. *Thinking about the Earth: A History of Ideas in Geology*. Cambridge, MA: Harvard University Press.

Ouspensky, P. D. 1912. *Tertium Organum: A Key to the Enigmas of the World*. New York: Vintage.

Paley, W. (1802) 1819. *Natural Theology*. Vol. 4 of *Collected Works*. London: Rivington.

Parascandola, J. 1971. Organismic and holistic concepts in the thought of L. J. Henderson. *Journal of the History of Biology* 4:63–113.

Peirce, C. S. 1934. *Scientific Metaphysics*. Vol. 6 of *Collected Papers of Charles Sanders Peirce*. Edited by C. Hartshorne and P. Weiss. 8 vols. Cambridge, MA: Harvard University Press.

Pennock, R., and M. Ruse, eds. 2008. *But Is It Science? The Philosophical Question in the Creation/Evolution Controversy*. 2nd ed. Buffalo, NY: Prometheus.

Pfeiffer, E. E. 1947. *The Earth's Face: Landscape and Its Relation to the Health of the Soil*. London: Faber and Faber.

———. 1958. Do we know what we are doing? DDT spray programs—their values and dangers. *Bio-Dynamics* 45:2–40.

Plotinus. 1992. *The Enneads: A New, Definitive Edition with Comparisons to Other Translations on Hundreds of Key Passages*. Translated and edited by S. Mackenna. Burdett, NY: Larson.

Podendorf, I. 1971. *Every Day Is Earth Day*. Chicago: Children's Press.

Postgate, J. 1988. Gaia gets too big for her boots. *New Scientist*, April 7, 80.

Prince of Wales, T. Juniper, and I. Skelly. 2010. *Harmony: A New Way of Looking at Our World*. New York: HarperCollins.

Razak, A. 1990. Toward a womanist analysis of birth. In *Reweaving the World: The Emergence of Ecofeminism*, edited by I. Diamond and G. F. Orenstein, 165–72. San Francisco: Sierra Club Books.

Repcheck, J. 2003. *The Man Who Found Time: James Hutton and the Discovery of the Earth's Antiquity*. London: Perseus.

Richards, R. J. 1987. *Darwin and the Emergence of Evolutionary Theories of Mind and Behavior*. Chicago: University of Chicago Press.

———. 2003. *The Romantic Conception of Life: Science and Philosophy in the Age of Goethe*. Chicago: University of Chicago Press.

———. 2008. *The Tragic Sense of Life: Ernst Haeckel and the Struggle over Evolutionary Thought*. Chicago: University of Chicago Press.

Rudwick, M. J. S. 1972. *The Meaning of Fossils*. New York: Science History Publications.

———. 2005. *Bursting the Limits of Time*. Chicago: University of Chicago Press.

Ruse, M. 1979. *Sociobiology: Sense or Nonsense?* Dordrecht, Holland: Reidel.

———. 1980. Charles Darwin and group selection. *Annals of Science* 37: 615–30.

———. 1982. *Darwinism Defended: A Guide to the Evolution Controversies*. Reading, MA: Benjamin/Cummings.

————, ed. 1988. *But Is It Science? The Philosophical Question in the Creation/Evolution Controversy*. Buffalo, NY: Prometheus.

————. 1995. Struggle for the soul of science. Review of P. Gross and N. Levitt, *Higher Superstition: The Academic Left and Its Quarrels with Science*. *The Sciences* (November/December): 39–44.

————. 1996. *Monad to Man: The Concept of Progress in Evolutionary Biology*. Cambridge, MA: Harvard University Press.

————. 1999a. *The Darwinian Revolution: Science Red in Tooth and Claw*. 2nd ed. Chicago: University of Chicago Press.

————. 1999b. *Mystery of Mysteries: Is Evolution a Social Construction?* Cambridge, MA: Harvard University Press.

————. 2003. *Darwin and Design: Does Evolution have a Purpose?* Cambridge, MA: Harvard University Press.

————. 2004. Adaptive landscapes and dynamic equilibrium: The Spencerian contribution to twentieth-century American evolutionary biology. In *Darwinian Heresies*, edited by A. Lustig, R. J. Richards, and M. Ruse, 131–50. Cambridge: Cambridge University Press.

————. 2005. Darwin and mechanism: Metaphor in science. *Studies in History and Philosophy of Biology and Biomedical Sciences* 36:285–302.

————. 2006. *Darwinism and Its Discontents*. Cambridge: Cambridge University Press.

————. 2008. *Charles Darwin*. Oxford: Blackwell.

————. 2010. Evolution and progress. In *Biology and Ideology from Descartes to Dawkins*, edited by D. Alexander and R. L. Numbers. Chicago: University of Chicago Press.

————, ed. 2013a. *The Cambridge Encyclopedia of Darwin and Evolutionary Thought*. Cambridge: Cambridge University Press.

————. 2013b. Evolution: From pseudo science to popular science, from popular science to professional science. In *Philosophy of Pseudoscience*, edited by M. Pigliucci and M. Boudry. Chicago: University of Chicago Press.

Sagan, L. 1967. On the origin of mitosing cells. *Journal of Theoretical Biology* 14:193–225.

Schelling, F. W. J. (1797) 1988. *Ideas for a Philosophy of Nature—as Introduction to the Study of This Science 1797—second edition 1803*. Translated by E. E. Harris and P. Heath. Cambridge: Cambridge University Press.

————. 2008. Timaeus. *Epoché* 12:205–48.

Schneider, S. H., and P. J. Boston, eds. 1991. *Scientists on Gaia*. Cambridge, MA: MIT Press.

Secord, J. A. 2000. *Victorian Sensation: The Extraordinary Publication, Reception, and Secret Authorship of "Vestiges of the Natural History of Creation."* Chicago: University of Chicago Press.

Sedgwick, A. 1845. Vestiges. *Edinburgh Review* 82:1–85.

———. 1850. *Discourse on the Studies at the University of Cambridge.* 5th ed. Cambridge: Cambridge University Press.

Sedley, D. 2008. *Creationism and Its Critics in Antiquity.* Berkeley and Los Angeles: University of California Press.

Segerstrale, U. 1986. Colleagues in conflict: An in vitro analysis of the socio-biology debate. *Biology and Philosophy* 1:53–88.

———. 2000. *Defenders of the Truth: The Battle for Science in the Sociobiology Debate and Beyond.* New York: Oxford University Press.

Sepkoski, D., and M. Ruse, eds. 2009. *The Paleobiological Revolution.* Chicago: University of Chicago Press.

Shapin, S. 1975. Phrenological knowledge and the social structure of early nineteenth-century Edinburgh. *Annals of Science* 32:219–43.

Sheldrake, R. 1981. *A New Science of Life: The Hypothesis of Morphic Resonance.* Los Angeles: J. P. Tarcher.

———. 1991. *The Rebirth of Nature: The Greening of Science and God.* New York: Bantam.

Shermer, M., ed. 2002. *The Skeptic Encyclopedia of Pseudoscience.* Santa Barbara, CA: ABC-CLIO.

Shuttleworth, J. 1975. The plowboy interview, part 2. *Mother Earth News* 32:6–17.

Sloan, P. R. 2012. Molecularizing Chicago, 1945–1965: The rise, fall, and rise of the biophysics program. Unpublished paper.

Smith, A. 1976. *The Glasgow Edition of the Works and Correspondence of Adam Smith.* Edited by R. H. Cambell and A. S. Skinner. Oxford: Clarendon Press.

Smuts, J. C. 1926. *Holism and Evolution.* London: Macmillan.

Sober, E., and D. S. Wilson. 1997. *Unto Others: The Evolution of Altruism.* Cambridge, MA: Harvard University Press.

Spencer, H. 1862. *First Principles.* London: Williams and Norgate.

———. 1868. *Essays: Scientific, Political, and Speculative.* London: Williams and Norgate.

Spretnak, C. 1989. Toward an ecofeminist spirituality. In *Healing the Wounds: The Promise of Ecofeminism,* edited by J. Plant, 127–32. Philadelphia, PA: New Society.

Starhawk. 1979. *The Spiral Dance: A Rebirth of the Ancient Religion of the Great Goddess.* Special twentieth anniversary edition. New York: Harper One.

———. 1990. Power, authority, and mystery: Ecofeminism and Earth-based spirituality. In *Reweaving the World: The Emergence of Ecofeminism,* edited by I. Diamond and G. F. Orenstein, 73–86. San Francisco: Sierra Club.

Steiner, R. 1914a. *Three Essays on Haeckel and Karma.* London: Theosophical Publishing House.

———. 1914b. *An Outline of Esoteric Science.* Hudson, NY: Anthroposophic Press.

———. 1924. *Agricultural Course: Birth of the Biodynamic Method.* Forrest Row, Sussex, UK: Rudolf Steiner Press.

———. 1957. *Karmic Relationships: Esoteric Studies.* Vol. 4. Translated by G. Adams, D. S. Osmond, and C. Davy. London: Anthroposophical Publishing Company.

Thoreau, H. D. 1854. *Walden.* Boston: Houghton, Mifflin.

———. 1864. *Maine Woods.* Boston: Tickner and Fields.

Turney, J. 2003. *Lovelock and Gaia: Signs of Life.* Cambridge: Icon Books.

Van Dyke, F., D. C. Mahan, J. K. Sheldon, and R. H. Brand. 1996. *Redeeming Creation: The Biblical Basis for Environmental Stewardship.* Downers Grove, IL: InterVarsity Press.

Vernadsky, I. 1998. *The Biosphere: Complete Annotated Edition.* New York: Springer.

Volk, T. 1998. *Gaia's Body: Toward a Physiology of the Earth.* New York: Copernicus Books.

Watson, A. J., and P. S. Liss. 1998. Marine biological controls on climate via the carbon and sulphur geochemical cycles. *Philosophical Transactions of the Royal Society,* series B, 353:41–51.

Watson, A. J., and J. E. Lovelock. 1983. Biological homeostasis of the global environment: The parable of Daisyworld. *Tellus, Series B: Chemical and Physical Meteorology* 35:284–89.

Wegener, A. 1915. *The Origin of Continents and Oceans.* Mineola, NY: Dover.

Wheeler, W. M. 1910. *Ants: Their Structure, Development and Behavior.* New York: Columbia University Press.

———. 1928. *The Social Insects: Their Origin and Evolution.* New York: Harcourt, Brace.

———. 1939. *Essays in Philosophical Biology.* Cambridge, MA: Harvard University Press.

Whewell, W. 1837. *The History of the Inductive Sciences.* 3 vols. London: Parker.

———. 1840. *The Philosophy of the Inductive Sciences.* 2 vols. London: Parker.

———. 1845. *Indications of the Creator.* London: Parker.

White, L. 1967. The historical roots of our ecological crisis. *Science* 155: 1203–7.

Williams, G. C. 1966. *Adaptation and Natural Selection.* Princeton, NJ: Princeton University Press.

Williamson, D. 2009. Caterpillars evolved from onychophorans by hybridogenesis. *Proceedings of the National Academy of Sciences* 106: 15786–90.

Wilson, D. S. 1980. *The Natural Selection of Populations and Communities.* Boston: Addison-Wesley.

Wilson, D. S., and E. O. Wilson. 2007. Rethinking the theoretical foundation of sociobiology. *Quarterly Review of Biology* 82:327–48.

Wilson, E. O. 1975. *Sociobiology: The New Synthesis.* Cambridge, MA: Harvard University Press.

———. 1984. *Biophilia.* Cambridge, MA: Harvard University Press.

———. 2002. *The Future of Life.* New York: Vintage Books.

Worster, D. 2008. *A Passion for Nature: The Life of John Muir.* Oxford: Oxford University Press.

Wright, S. 1931. Evolution in Mendelian populations. *Genetics* 16:97–159.

———. 1932. The roles of mutation, inbreeding, crossbreeding and selection in evolution. *Proceedings of the Sixth International Congress of Genetics* 1:356–66.

———. 1945. Tempo and mode in evolution: A critical review. *Ecology* 26:415–19.

Young, R. M. 1985. *Darwin's Metaphor: Nature's Place in Victorian Culture.* Cambridge, MA: Cambridge University Press.

Zell-Ravenheart, O. 2004. *Grimoire for the Apprentice Wizard.* Franklin Lakes, NJ: New Page Books.

———. 2006. *Companion for the Apprentice Wizard.* Franklin Lakes, NJ: New Page Books.

———. 2009. *Green Egg Omelette: An Anthology of Art and Articles from the Legendary Pagan Journal.* Franklin Lakes, NJ: New Page Books.

Index